Assessing the Training and Operational Proficiency of China's Aerospace Forces

Selections from the Inaugural Conference of the China Aerospace Studies Institute (CASI)

Edmund J. Burke, Astrid Stuth Cevallos, Mark R. Cozad, Timothy R. Heath

Prepared for the United States Air Force
Approved for public release; distribution unlimited

For more information on this publication, visit www.rand.org/t/CF340

Library of Congress Cataloging-in-Publication Data is available for this publication.
ISBN: 978-0-8330-9549-7

Published by the RAND Corporation, Santa Monica, Calif.

Preface

On June 22, 2015, the China Aerospace Studies Institute (CASI), in conjunction with Headquarters, Air Force, held a day-long conference in Arlington, Virginia, titled "Assessing Chinese Aerospace Training and Operational Competence." The purpose of the conference was to share the results of nine months of research and analysis by RAND researchers and to expose their work to critical review by experts and operators knowledgeable about U.S. and Chinese air, space, and missile operations. The research drew heavily on openly published Chinese language sources, many of which have not been translated into English.

CASI researchers presented eight papers at the conference and received valuable feedback from discussants and participants, both during the conference and afterward. Revised versions of three of the papers presented at the conference are collected in this volume:

- "PRC Leadership Dynamics Shaping the Future of the PLA Air Force: Examining the Party's Military Reform Efforts," Edmund J. Burke and Astrid Stuth Cevallos
- "Trends in PLA Air Force Joint Training: Assessing Progress in Integrated Joint Operations," Mark R. Cozad and Astrid Stuth Cevallos
- "New Type Support: Developments in PLAAF Air Station Logistics and Maintenance," Timothy R. Heath.

Three other papers presented at the conference have been published separately by RAND Project AIR FORCE:

- *Overview of People's Liberation Army Air Force "Elite Pilots,"* Michael S. Chase, Kenneth W. Allen, and Benjamin S. Purser III, RR-1416-AF, 2016
- *Training the People's Liberation Army Air Force Surface-to-Air Missile (SAM) Forces,* Bonny Lin and Cristina L. Garafola, RR-1414-AF, 2016
- *From Theory to Practice: People's Liberation Army Air Force Aviation Training at the Operational Unit,* Lyle J. Morris and Eric Heginbotham, RR-1415-AF, 2016.

Finally, two papers are being published by the Jamestown Foundation and are available at www.jamestown.org:

- "PLA Air Force Aviator Recruitment, Education, and Training," Kenneth W. Allen
- "Building a Strong Informatized Strategic Missile Force: Overview of the Second Artillery Force with a Focus on Training in 2014," Kenneth W. Allen and Jana Allen.

About the China Aerospace Studies Institute

CASI was created in 2014 at the initiative of Headquarters, Air Force. CASI is part of the RAND Corporation's Project AIR FORCE (PAF); Air University and Headquarters Pacific Air Forces (PACAF) are key stakeholders. The purpose of CASI is to advance understanding of the capabilities, operating concepts, and limitations of China's aerospace forces. Its research focuses on the People's Liberation Army (PLA) Air Force (PLAAF), Naval Aviation, Second Artillery, and the Chinese military's use of space. CASI aims to fill a niche in the China research community by providing high-quality, unclassified research on Chinese aerospace developments in the context of U.S. strategic imperatives in the Asia-Pacific region. CASI will transition to an independent, Air Force–supported organization in fiscal year 2017, with personnel at Fort Lesley J. McNair in Washington, D.C., and at Air University, Maxwell Air Force Base, Alabama.

More than two decades ago, the armed forces of the People's Republic of China embarked on a series of efforts to transform themselves from a massive, low-tech force focused on territorial defense to a leaner, high-tech force capable of projecting power and influence around the peripheries of the nation, even against the expeditionary forces of the most capable adversary. By developing operational strategies tailored to meeting this objective; reforming the organization, training, and doctrine of the armed forces; and, perhaps most importantly, making large and sustained investments in new classes of weapon systems, China may be on the cusp of realizing the ambitious goals it has laid out for its armed forces. Without doubt, China's air and maritime forces today field capabilities that are compelling U.S. military planners to rethink their approach to power projection and to reorient important components of their modernization programs. China, in short, has become the "pacing threat" for the United States Air Force and Navy. Therefore, the importance of understanding the capabilities and limitations of China's air, naval, and space forces is self-evident. CASI's mission is to contribute to that understanding.

CASI's research team brings to this work a mastery of research methods, understanding of China's military capabilities and doctrine, and the ability to read and understand Chinese writings. When undertaking research for CASI reports, analysts used a variety of Chinese-language primary source documents on PLA and PLAAF training, operations, and doctrine. This includes *Kongjun Bao* (Air Force News) and *Huojianbing Bao* (Rocket Force News)—the daily newspapers of the PLAAF and PLA strategic missile forces—as well as defense white papers, PLA encyclopedias, and books written by military officers and academics affiliated with the PLA, such as the Academy of Military Science. These publications are considered authoritative assessments and reporting on training, strategy, and concepts for how the PLAAF and missile forces prepare for military operations and warfare in general. It is important to acknowledge, however, that these PLA publications also have some weaknesses and that reliance on open sources necessarily has some limitations. The growing availability of primary source material helps to compensate for at least some of these challenges.

Additional information about CASI is available on RAND's CASI website:
www/rand.org/paf/casi

RAND Project AIR FORCE

RAND Project AIR FORCE (PAF), a division of the RAND Corporation, is the U.S. Air Force's federally funded research and development center for studies and analyses. PAF provides the Air Force with independent analyses of policy alternatives affecting the development, employment, combat readiness, and support of current and future air, space, and cyber forces. Research is conducted in four programs: Force Modernization and Employment; Manpower, Personnel, and Training; Resource Management; and Strategy and Doctrine. The research reported here was prepared under contract FA7014-06-C-0001.

Additional information about PAF is available on our website:
www.rand.org/paf/

Contents

Chapter One

PRC Leadership Dynamics Shaping the Future of the PLA Air Force

Examining the Party's Military Reform Efforts

Edmund J. Burke and Astrid Stuth Cevallos

Summary

This report examines the impact that Xi Jinping's priorities and the composition of the Central Military Commission (CMC) might have on People's Liberation Army (PLA) Air Force (PLAAF) influence, bureaucratic leverage, and force development goals in the near and long term. Xi has consolidated his control of state, party, and military bodies while overseeing an anticorruption campaign that is far more extensive than many would have thought possible when he took power in 2012. On the operational side, new mission areas—in the maritime domain, in particular—are driving demands for higher readiness rates, while more-realistic training, under what press reports term "actual combat" or "wartime" conditions, is challenging PLAAF commander Ma Xiaotian and his PLAAF leadership team to move more quickly on a host of initiatives, most importantly a new training regimen to prepare the PLAAF for China's version of joint operations with its PLA Navy (PLAN), PLA Army (PLAA), PLA Rocket Forces (PLARF), and Strategic Support Forces counterparts.

The appointment of former PLAAF commander Xu Qiliang to a vice chairman post on the CMC could signal that the PLAAF's bureaucratic heft is increasing and that PLAAF issues and funding will receive a higher priority in coming years. In the near term, the anticorruption campaign is high on Xu's agenda, as he is in charge of the CMC inspection work leading group and therefore is the senior officer in charge of Xi's blanket inspection of the officers' corps. But the cultural problems signaled by the widespread anticorruption campaign are longstanding and have proven resistant to change since the mid-1990s. The difficulties Xu will face during his tenure were magnified in early 2016, when sweeping military reforms were announced; success in implementing these reforms by 2020 will in part shape Xu's legacy in the short run. In the longer term, Xu and his successor, Ma, are responsible for laying a foundation to meet ambitious force development goals out to 2030 and beyond, by which time the PLAAF hopes to be a powerful force for integrated air and space operations, capable of attack and defense.

Introduction

Since ascending to the top spots in China's government, party, and military in 2012, Xi Jinping—general secretary of the Chinese Communist Party (CCP), president of the People's Republic of China (PRC), and chairman of the CMC, China's highest military deliberative and decisionmaking body—has presided over the largest anticorruption campaign since the days of Chairman Mao Zedong.[1] At the same time, he has called meetings where he elicited expressions of fealty from military commanders and reiterated Mao's edict that the "Party commands the gun." Finally, the past few years have witnessed the unprecedented promotion of a PLAAF commander to vice chairman of the CMC, resulting in the presence of not one, but two PLAAF generals on the CMC (the other being PLAAF commander Ma Xiaotian). What do all of these moves among military leadership mean for the future of the PLAAF?

In considering that question, this report begins with an examination of two stage-setting and unusual political events: an April 2014 meeting where senior PLA leaders pledged their allegiance to Xi and a November 2014 reprisal of the historic 1929 Gutian Conference, where Xi reinforced the CCP's control over the PLA. Against the backdrop of Xi's anticorruption campaign, these events paint a picture of Xi's efforts not just to enhance the CCP's political legitimacy but also to consolidate his control over the military. The report then explores the paths to leadership of former PLAAF commander and current Vice Chairman of the CMC Xu Qiliang and current PLAAF commander and CMC member Ma Xiaotian before turning to a consideration of their potential legacies, especially with regard to achieving the PLAAF's (and PLA's) strategic goals and Xi's much-touted "China Dream." It concludes with an analysis of the possible long-term implications of Xu and Ma's leadership for the future of the PLAAF.

This report finds that the PLAAF has gained a measure of status and influence due to structural changes in the PLA and that Xu and Ma are the trailblazers for a new dynamic in PLAAF leadership.[2] Their presence on the CMC should help PLAAF make gains in the bureaucratic competition for resources, if only because the structural changes were designed to give services other than the PLAA a more prominent place in military leadership. The PLAAF may also achieve an expanded mission set as the PLA evolves into a more modern joint force. In addition, Xu's promotion to vice chairman of the CMC may not be a one-off occurrence, but might instead be a byproduct of these same structural changes that now have PLAAF officers attaining not just the rank but also the grade necessary to reach the highest levels of leadership in

[1] Shai Oster, "President Xi's Anti-Corruption Campaign Biggest Since Mao," *Bloomberg Business*, March 4, 2014.

[2] Though it has existed since 1949, the PLAAF was marginalized within the PLA following the Cultural Revolution and Lin Biao's death in the early 1970s. In the 1980s, the PLAAF began a process of comprehensive reform, but it remained marginalized through the mid-1990s, when the PLA began to modernize in earnest. For more on the early history of the PLAAF, see Kenneth W. Allen, Glenn Krumel, and Jonathan D. Pollack, *China's Air Force Enters the 21st Century,* Santa Monica, Calif.: RAND Corporation, MR-580-AF, 1995.

the PLA.[3] The Chinese military leadership and personnel system has never before been structured to allow PLAAF (or PLAN) leaders to attain the grade, rank, and position that would give them an opportunity to fill the top three positions in the PLA.

An Unusual Pledge of Allegiance and a Conference Affirming the Primacy of Party over Army

In early April 2014, PLAAF commander Ma Xiaotian took part in an extraordinary meeting: He and 17 other senior PLA leaders all but swore their personal allegiance to Xi Jinping.[4] In comments attributed to him under the heading "Strive to Enhance the Air Force's Ability to Fight and Win," Ma praised Xi for providing "the ideological and theoretic guide for the whole Party, the whole military, the entire people throughout the country to make concerted efforts for the fulfillment of the great rejuvenation of the Chinese nation at an earlier date."[5] His peers—fellow service heads, commanders of each of China's seven military regions, and senior officers from each of the PLA's four general departments—each also spoke and submitted articles, all of which were printed collectively on two full pages in the PLA's daily newspaper on April 3, 2014.[6] Each extolled Xi and his proper leadership and guidance, and then individually listed the pressing imperatives in terms of force development or organizational obligations for their respective services and entities. The English-language version of China's official newspaper covered the story under the title "PLA Senior Generals Back Xi's Orders."[7] Regional press outlets highlighted how unusual this show of fealty was but varied as to just exactly what it meant. The *South China Morning Post* declared it a sign of Xi's strength among the military,

---

[3] On the distinction between grades and ranks, Kenneth Allen writes, "In the PLA . . . grades are based on an officer's position and are more important than ranks. As a result, PLA writings usually refer to officer positions or grades and have few references to ranks. Within the PLA, an officer's grade, not rank, reflects authority and responsibility across service, branch, and organizational lines. While rank is a key indicator of position within the hierarchy of foreign militaries, grade is the key indicator of authority within the PLA." See Kenneth Allen, "Assessing the PLA's Promotion Ladder to CMC Member Based on Grades vs. Ranks—Part 1," *China Brief*, Vol. 10, No. 15, July 22, 2010a.

[4] "PLA Senior Generals Back Xi's Orders," *China Daily*, April 3, 2014.

[5] Ma Xiaotian [马晓天], "Strive to Enhance the Air Force's Ability to Fight and Win" [努力提高空军部队能打仗打胜仗能力], *Liberation Army Daily* [解放军报], April 2, 2014.

[6] Between late 2014 and early 2015, the PLA announced a major organizational reform to the structure of the Chinese armed forces, which included revamping the CMC, dissolving the four general departments (GDs), creating a new service headquarters, and establishing five new "theater commands" to replace the seven military regions. This chapter was written before the organizational reform was announced and therefore refers to the original GD structure. For more on the reform, see Kenneth Allen, Dennis J. Blasko, and John F. Corbett, "The PLA's New Organizational Structure: What is Known, Unknown and Speculation (Part 1)," *China Brief*, Vol. 16, No. 3, February 4, 2016; "In Depth Study and Implementation of Chairman Xi Jinping's Important Expositions on National Defense and Army Building, Pushing Forward the Great Practice of Building and Strengthening the Military From a New Starting Point" [深入学习贯彻习主席关于国防和军队建设重要论述，在新的起点上推进强军兴军伟大实践], *Liberation Army Daily* [解放军报], April 2, 2014.

[7] "PLA Senior Generals Back Xi's Orders," 2014.

while *The Straits Times* in Singapore cautioned that Xi might not be "facing any ordinary opposition but possibly a challenge to his political legitimacy."[8]

In retrospect, the *South China Morning Post* had it right. The April 2014 meeting was a watershed public moment that signaled Xi's consolidation of absolute power atop the PLA and the CCP after a tumultuous two years of infighting that saw the ousting of many high-ranking officials loyal to previous CCP leaders Hu Jintao and Jiang Zemin. The meeting was held just days after Gu Junshan, a former deputy director of the General Logistics Department (GLD), who had been removed from his position in February 2012,[9] was formally indicted in a Beijing military court on a series of corruption and embezzlement charges.[10] Gu's case marked the beginning of what has become Xi's anticorruption campaign in the military. Since 2013, over 4,000 officers at or above the rank of lieutenant colonel—including more than 80 generals—have been investigated; over 20 officers have been removed from their posts, and nearly 150 have been demoted.[11] In March 2014, the PLA published the names of 14 generals under investigation or already convicted for corruption, including Rear Admiral Guo Zhenggang, the son of former CMC Vice Chairman Guo Boxiong.[12]

Yet this unusual show of loyalty by Ma and his peers to China's top leader was soon to be upstaged. In November 2014, Xi convened a meeting of 400 senior officers in southeastern China's Fujian Province. Besides being the province where Xi spent some 17 years of his career, ascending from low-level postings through the CCP system, Fujian is also home to Gutian, a site steeped in CCP and army history. There, in 1929, the CCP met for the first time since the Red Army was founded in 1927—an event that came to be known as the Gutian Conference.[13] At the time, Mao (along with comrade He Long) was leading the Red Army, then a ragtag group of a few thousand impoverished, uneducated peasants who had survived several disasters and massacres in the countryside and were surviving with the support of the local population in rural

---

[8] Zhang Hong, "PLA Generals Take Rare Step of Swearing Loyalty to President Xi Jinping," *South China Morning Post Online*, April 3, 2014; Kor Kian Beng, "18 PLA Generals Show Collective Support for China's President Xi," *The Straits Times*, April 3, 2014.

[9] "Personnel Changes at the People's Liberation Army General Logistics Department: Gu Junshan Is No Longer Deputy Director" [解放军总后勤部人事调整，谷俊山不再担任副部长], *People's Daily* [人民日报], February 11, 2012.

[10] "Handle the Trial of Gu Junshan's Case Strictly on the Basis of Law – PLA Expert Yu Xiao Interprets Relevant Legal Issues in the Trial of Gu Junshan's Case" [严格依法审理谷俊山案件 — 军队法律专家欲晓解读审理谷俊山案件有关诉讼法律问题], *Liberation Army Daily* [解放军报], April 1, 2014.

[11] "Scores of PLA Officers Punished," *China Daily*, January 30, 2015.

[12] Dan Levin, "China Names 14 Generals Suspected of Corruption," *New York Times*, March 2, 2015.

[13] The Red Army—officially termed the First Workers' and Peasants' Army—was founded after the August 1927 Nanchang Uprising as the military wing of the CCP. As the Chinese civil war continued following Japan's defeat in World War II, the Red Army was renamed the People's Liberation Army.

of the Party and the PLA" and warned that "military reform has entered the stage of tackling tough issues and has come to a crucial juncture of style rectification and anticorruption."[21] Xi's speech clarified and underscored the central role of the antigraft drive in tightening the CCP's grip on the PLA. After Gutian, the GPD argued, the PLA not only "will permanently maintain the Party's absolute leadership over the armed forces, and will persistently act and fight under the Party's leadership," but also "must address the existing prominent problems in the military with the courage of scraping poisoned tissues off bones."[22]

The Anti-Corruption Campaign: Xi Jinping Dictates the Agenda

Xi Jinping's sweeping anticorruption campaign can be interpreted as a sign of his mastery of the Chinese political system since his rise to the top, as he has navigated a nearly uncharted path that has addressed a longstanding source of public anger and resentment toward the CCP. Indeed, according to the findings of the 2013 Pew Research Global Attitudes Project, 53 percent of Chinese feel that corrupt officials are a "very big problem" in their country, up from 50 percent in 2012 and 38 percent in 2008.[23] In a speech to the Politburo following his promotion to CCP chairman in November 2012, Xi noted how corruption has driven social unrest and regime change in other countries, concluding with the observation that "corruption could kill the Party and ruin the country."[24] Since then, Xi has made rooting out corruption a priority, targeting officials of all ranks and introducing austerity measures to reduce frivolous spending on lavish banquets and gifts. In what appears to be the farthest-reaching probe of government officials since the Mao era, Xi has ensnared "tigers" (high-ranking officials), "flies" (low-ranking officials), and "foxes" (officials of any rank who have fled abroad with ill-gotten gains). The most senior official toppled for corruption since the founding of the PRC, former Politburo Standing Committee member and Minister of Public Security Zhou Yongkang, was expelled from the CCP in December 2014 and subsequently sentenced to life in prison in June 2015 on charges of bribe-taking, adultery, and leaking state secrets. Between 2013 and early 2015, Xi's antigraft campaign entangled about 100,000 officials.[25]

In the Chinese system, it is never clear why some officials are sacked and expelled from the CCP for corruption allegations while others emerge unscathed, and there are differing

---

[21] PLA General Political Department [中国人民解放军总政治部], "The Scientific Guide to Strengthening and Developing the Military Under the New Situation—Deeply Study and Implement Chairman Xi Jinping's Important Speech at the All-PLA Political Work Conference" [新形势下强军兴军的科学指南 — 深入学习贯彻习近平主席在全军政治工作会议上的重要讲话], *Seeking Truth* [求是], December 1, 2014.

[22] PLA General Political Department, 2014.

[23] Pew Research Center, "Environmental Concerns on the Rise in China: Many Also Worried About Inflation, Inequality, Corruption," September 19, 2013.

[24] Zhao Yinan, "Xi Repeats Anti-Graft Message to Top Leaders," *China Daily*, November 20, 2012.

[25] Aibing Guo, "China's Top Anti-Graft Regulator to Inspect 26 State Companies," *Bloomberg Business*, February 11, 2015.

perspectives on this subject. Some posit that the anticorruption campaign is more about promoting good governance and strengthening the Party than about weakening opposing forces within the CCP who could threaten Xi's rule. Others emphasize that power politics are always in play, and that Xi is merely using the anticorruption campaign to achieve his own political goals. The truth likely lies somewhere in the middle, as Xi's vision of good governance appears to be driven by the survival and consolidation of the CCP regime under the control of a second generation of "princelings"—those with familial ties to the founding generation of CCP military and political leaders. Indeed, many from this second generation—including Xi himself—now populate the upper echelons of military and political leadership.

Methodically dismantling entrenched networks of corrupt senior officials and cronyism has a way of activating opposition and vested interest groups within that system, making the prospect of an eventual challenge to Xi's Party leadership plausible. At this point, however, no direct threats to Xi appear on the horizon, insofar as we can determine from open-source materials, and by all accounts, he is at the height of his political powers. The emphasis on politics and the CCP in the military mirrors an all-out CCP campaign to create in Xi the personification of the Chinese state and CCP that has been under way since the beginning of his time at the top of the Chinese leadership. Coverage of Xi and his exploits are everywhere in China in a way that is a sharp departure from the comparatively boring public profile of his predecessor, Hu Jintao.[26] Xi's influence over official media appears to be stronger, too: In February 2016, as Xi embarked on a tour of state media outlets to emphasize the news media's primary role as the Party's mouthpiece, the staff of China Central Television, *China Daily*, and Xinhua pledged their loyalty to him—not unlike the PLA generals who did so in April and November 2014.[27]

Arguably, Xi is actively creating a new direction for the Party and its military. An increasing sense of nationalism, embodied in China's growing assertiveness in the East and South China Seas, is part and parcel of the "China Dream," which has been described as Xi's political ideology. While specific definitions are fluid, it has been concisely described as "national rejuvenation, improvement of people's livelihoods, prosperity, construction of a better society and military strengthening . . . that can be best achieved under one party, Socialist rule."[28] The China Dream has been invoked repeatedly to express the ideal and image of a strong China assuming its rightful place on the world stage, and it both reflects and stokes a streak of nationalism among the populace that certainly includes a strong military befitting a rising power. Former PLAAF commander and current CMC Vice Chairman Xu Qiliang and his successor, Ma Xiaotian, face the challenge of making Xi's China Dream and "strong army" dream a reality by transforming the PLAAF and its role in PLA operations. They represent the vanguard of a new

---

[26] Evan Osnos, "Born Red," *The New Yorker*, April 6, 2015.

[27] Edward Wong, "Xi Jinping's News Alert: Chinese Media Must Serve the Party," *New York Times*, February 22, 2016; Xinhua, "President Xi Visits China's National News Outlets," February 19, 2016.

[28] Bill Bishop, "A Highly Public Trip for China's President, and Its First Lady," *New York Times*, March 25, 2013.

path to PLAAF leadership enabled by significant structural and personnel reforms dating back roughly a decade that were designed to lessen the PLA's domination of senior leadership positions and its hammerlock on military culture.

To the extent that one exists at all, the case Xi is making about his China Dream certainly could appeal to a nascent sense of duty and service among the officer corps, but the cultural problems are longstanding. Political leaders have delivered on material improvements and substantial defense budget increases since the mid-1990s and likely expected more than rampant, systemic corruption from their senior officers in return. Like all other PLA senior officers, PLAAF leaders were, at the least, witting of this systemic problem and therefore were part of the problem by definition. They now also face the daunting task of managing the most extensive military reform initiatives in at least 30 years as the PLA downsizes and reorganizes for joint operations, with a target date of 2020.[29]

Xu Qiliang and Ma Xiaotian's Paths to Leadership

Xu Qiliang: A Transformative Figure?

With his appointment in November 2012, General Xu Qiliang made PLA history by becoming the first PLAAF officer to become a vice chairman of the CMC.[30] (China has no equivalent to the U.S. Joint Chiefs of Staff, and military leadership has been thoroughly dominated by ground force officers throughout its history.) It is unclear if Xi Jinping, who had just assumed the CMC chairmanship, had control over his selection, but Xi appears to be able to work with (and through) Xu. Xu has achieved many other "firsts" in his career: He was the first vice chairman born in the 1950s[31] and was the first PLAAF officer to serve as a deputy chief of the PLA General Staff Department (DCGS) when he was appointed in July 2004, immediately preceding his service as PLAAF commander.[32] Perhaps more important than any other position or experience on his resume is his participation in the PLA's military deployment to the Vietnam border some 35 years ago. In 1979, Xu was the regimental chief of staff with the 40th Regiment, 14th Air Division on the Taiwan Strait. A generation earlier, the unit had fought against Nationalist forces, including in the artillery bombardment of Jinmen in 1958. Given its mission and basing on the Taiwan Strait, it was often in combat readiness status and was considered an elite unit. The regiment, along with many of China's other frontline air units, deployed near

---

[29] Reuters, "China's Xi Sets Up Five New 'Battle Zones' in Military Reform Push," February 1, 2016.

[30] Ma Hao-Liang [马浩亮], "New Records in the Promotion of Central Military Commission Vice Chairmen" [军委副主席普生创纪录], *Ta Kung Pao* [大公报], November 5, 2012.

[31] Ma Hao-Liang, 2012.

[32] Chin Chien-Li [金千里], "One of the Key People in Chinese Communist Combat Operations Against Taiwan: Critical Biography of the General Staff Lieutenant General Xu Qiliang" [中共对台作战中坚人物： 中共副总参谋长许其亮中将评传], *The Frontier* [前线], June 1, 2005.

Vietnam but never crossed the border when the two sides entered conflict on the ground in February 1979.[33] While neither Xu's unit nor any other PLAAF unit actually engaged Vietnamese forces in combat, PLAAF units maintained air superiority in an environment close to real combat. As a result, Xu is part of a very small group of PLAAF officers who were at least in a war zone and can therefore both draw on that experience and use it to their advantage when vying against peers who lack the same credential. In short, it is a formative experience that further distinguishes from his peers an officer already positioned to succeed.

The selection of the other vice chairman of the CMC in 2012 also broke new ground, meaning that both of the new vice chairmen attained their positions as a result of the structural reforms to the PLA's promotion process. PLAA General Fan Changlong became the first PLA general ever appointed to the position without having previously served on the CMC. Fan's résumé for making this unprecedented jump was burnished in his most recent post as Jinan MR commander, where he was given credit for overseeing a pilot project to develop and implement a joint logistics system across the services in the region, as well as a series of joint and multinational exercises.[34] His rise without first attaining the CMC Member grade meant that he leapfrogged over Xu and Navy commander Wu Shengli for the first vice chair position; Fan is now listed as first in the PLA in protocol order. This jump gave Fan the operational portfolio on CMC, while Xu, in the second vice chairman slot, has the political commissar portfolio.

Xu and Fan's selections as vice chairmen have also been assessed through the prism of Chinese factional leadership politics. Some reports link Xu to former leader Jiang Zemin, who chaired the CMC from 1989 to 2004, and Fan sometimes is described as an ally of Hu Jintao, who succeeded Jiang and transferred power to Xi in 2012.[35] While Fan spent the bulk of his career in northeast China's Shenyang MR, which has been referred to as the "northeast army faction"[36] headed by Xu Caihou—the highest-ranking former military official felled in the current anticorruption campaign so far and the highest-ranking to be ensnared in some 20 years—it appears that he did not have strong personal ties to Xu.[37] Fan has publicly aligned himself in official speeches as supportive of Xu Caihou's ousting[38] and no doubt could not have skipped CMC Member grade and risen to his post without Xi Jinping being comfortable with his

---

[33] Chin Chien-Li, 2005.

[34] Chin Chien-Li, 2005.

[35] Choi Chi-yuk, "PLA Leadership Changes Reflect a Fine Balancing Act," *South China Morning Post*, October 26, 2012b.

[36] Choi Chi-yuk, "General Fan Changlong Tipped for Top Post in China Military Commission," *South China Morning Post*, October 22, 2012a.

[37] Angela Meng, "PLA Reshuffle Strengthens Xi Jinping's Hand in Corruption Fight," *South China Morning Post*, September 22, 2014; Ma Cheng-kun, "Assessing the New PLA Leadership," presentation at CAPS-RAND-NDU Conference on the PLA, November 2014.

[38] Xinhua, "Fan Changlong Conducts Fact-Finding at Tibet- and Qinghai-Based Army Units, Emphasizes Need to Deeply, Solidly Conduct Live Combat Military Training, Constantly Increase Army Units' Ability to Perform Their Functions and Missions" [范长龙在驻西藏、青海部队调研时强调深入扎实抓好实战化军事训练，不断提高部队履行职能任务能力], August 17, 2014a.

allegiances. In reality, the simple math of the PLA's rank and grade structure appears to have presented Chinese military and political leadership with the need to elevate either PLAAF officer Xu Qiliang or PLAN officer Wu to the highest military position on the CMC ahead of an Army officer. They chose instead to elevate Fan, perhaps because of his experience with developing joint operational capabilities in Jinan MR, but almost certainly also because he represents PLA ground forces—and therefore continuity.[39]

As the vice chairman with the political commissar portfolio, Xu is the front man of the anticorruption drive in the military, a very delicate—and potentially treacherous—assignment that likely dominated his attention from early 2014 through mid-2015. Xu presided over the first meeting of the CMC's newly established inspection work leading group in late October 2013,[40] and in late November 2013 presided over a training session and meeting for inspection team members.[41] In November 2014, likely as a finding of the CMC working group chaired by Xu, the duties of auditing PLA officials and organs were taken away from the GLD and moved directly under the CMC. A January 2015 summary of the audit progress since the CMC took over the function stated that over 4,000 officers at regiment and higher levels had been audited, 21 had been dismissed, and 144 had their posts "adjusted"—presumably meaning they were demoted.[42] By April 2015, when the heads of both the Lanzhou and Beijing MR Joint Logistic Departments were detained, five of the seven regional commands had lost their senior logistics commanders to corruption charges.[43]

While important in assessing Xu's role and influence within the CMC, this political commissar portfolio does not give Xu direct control over PLAAF operational issues or procurement; both of those concerns fall under Fan's portfolio as the operational chief of the

---

[39] Ma Cheng-kun agrees with this analysis, arguing that Fan was chosen because (a) he was capable and had performed well as Jinan MR commander (the smallest and least important MR, according to Ma); (b) he did not have a close personal connection to either Guo Boxiong or Xu Caihou, the two previous CMC vice chairmen; (c) he is an army general, and at least one CMC vice chairman has traditionally represented the ground forces; and (d) Fan should not have expected an additional promotion after reaching the position of Jinan MR commander, and therefore his unexpected promotion to CMC Vice Chairman at Xi's request makes him indebted to Xi. Thus, Ma concludes, Xi's selection of Fan as CMC vice chairman revealed the importance Xi placed on continuing the anticorruption campaign and rooting out the influence of PLA "senior brass." Ma Cheng-kun, 2014.

[40] Yang Ting [杨婷], ed., "With the Approval of Xi Jinping, the Central Military Commission Prints and Issues the 'Decision of the Central Military Commission on Unfolding Inspection Work' and the 'Regulations of the Central Military Commission on Inspection Work (Trial)'; Xu Qiliang Chairs a Meeting to Make Implementation Arrangements" [经习近平主席批准中央军委引发《中央军委关于开展巡视工作的决定》《中央军委巡视工作规定（试行），许其亮主持召开会议对贯彻落实作出部署], Xinhua, October 29, 2013.

[41] "PLA Inspection Tour Organs Established, PRC Ministry of National Defense," *China Military Online*, November 22, 2013.

[42] Xie Shenzong [解审综] and Sai Zongbao [塞宗宝], "Pushing Forward Innovative Development of Armed Forces' Audit Work from New Starting Point—Roundup on All Army and Armed Police Units Focusing on Strong-Military Objective To Strengthen Audit Work" [在新起点上推动军队审计工作创新发展—全军和武警部队聚集强军目标加强审计工作综述], *Liberation Army Daily* [解放军报], January 29, 2015.

[43] Reuters, "China Investigates Three Senior Army Officers Suspected of Graft," April 27, 2015.

PLA. Xu's position instead allows him to influence collective decisionmaking and to advocate for the PLAAF behind closed doors.

Ma Xiaotian: Following in Xu's Footsteps?

For Ma Xiaotian, commanding the PLAAF and receiving that position's associated CMC membership are the crowning achievements of a career that, like those of some of his peers, was boosted by heredity and familial ties: His father, Senior Colonel Ma Zaiyao, once served as the dean of students of the PLA Political College, while his father-in-law, Lieutenant General Zhang Shaohua, was the deputy secretary of the CMC's Discipline Inspection Commission. Today, inclusion in this "princeling" set is the surest ticket to leadership positions throughout the Chinese political and military systems.[44] Such familial ties are likely even more important in a military service that has gone over 35 years without the types of combat experience that might allow pilots to distinguish themselves among their peers.

Ma joined the PLAAF at age 16—like his predecessor Xu Qiliang—at the beginning of the Cultural Revolution in 1965 and at one point became the youngest deputy regiment commander in the PLAAF. He successively progressed through operational units until his appointment as one of the PLAAF deputy chiefs of staff in 1997. Over the subsequent nine years, Ma held one of the top three positions in each of three regional commands (Guangzhou, Lanzhou, and Nanjing Military Region Air Force commands) and then was appointed as one of the PLAAF's deputy commanders. He is also credited with having piloted a Su-30 multirole fighter at the age of 49 at the Zhuhai Air Show in 1998.[45] From 2006 to 2007, Ma was the commandant of the National Defense University, China's premier military educational institution, becoming the first PLAAF general to serve in this position.[46] Finally, from 2007 to 2012, immediately prior to assuming command of the PLAAF, Ma served as one of four DCGS, becoming just the second PLAAF officer to hold this post. (As previously noted, Xu became the first PLAAF officer to serve as a DCGS in 2004.) It appears that Xu and Ma are establishing a precedent—that PLAAF officers' path to PLAAF commandership and CMC membership necessarily involves serving as DCGS to make grade to serve at the highest levels. Unlike PLAA officers, who attain the grade necessary for promotion to commander of the PLAA by serving as one of seven MR commanders, PLAAF and PLAN officers are not eligible to serve as MR commanders. Therefore, at the moment, promotion to one of the four DCGS positions appears to be the only path that can allow PLAAF

---

[44] For an excellent discussion of familial ties as they relate to military and leadership dynamics in general, see Cheng Li, "China's Midterm Jockeying: Gearing Up for 2012 (Part 3: Military Leaders)," *China Leadership Monitor*, No. 33, 2010.

[45] Minnie Chan, "PLA Hawk Ma Xiaotian Flying Ever Higher," *South China Morning Post*, October 25, 2012.

[46] Chin Chien-Li [金千里], "The CPC's Key Figure in Conflict with Taiwan: A Critical Biography of Lieutenant General Zhang Qinsheng, the New Deputy Chief of General Staff—Prefect of Hu Jintao's 'Class of Military Generals'" [中共对台作战中坚人物：新任副总参谋长章沁生中将评传—胡锦涛的'将军版'班长], *The Frontline* [前线], March 1, 2007.

officers to attain the grade necessary for subsequent promotion to service commander. This means that we should probably look at PLAAF officers who are promoted to the DCGS position as the only candidates eligible from a grade perspective to take on the mantle of PLAAF commandership in the future.[47]

The DCGS position in particular gave Ma wide exposure to foreign government and military delegations, as representing the PLA in foreign affairs was part of his portfolio. As a rough equivalent to a deputy secretary of defense in the U.S. system, Ma's official duties also ensured he would be quoted extensively while speaking officially on the PLA's behalf to international audiences. His tenure coincided with a tense time in U.S.–Chinese military-to-military relations, which were plagued by Chinese concerns surrounding Taiwan's presidential election in March 2008 and nagging incidents like the flight test of the J-20 fighter during a 2010 visit to Beijing by then–Secretary of Defense Robert Gates. Ma proved up to the task of delivering both stern and conciliatory talking points, depending on the situation. For example, while heading the Chinese delegation to Washington for the ninth Defense Consultative Talks in December 2007, he delivered China's perspective on a Taiwan independence referendum and urged the U.S. side to stop its military contacts with Taiwan, including arms sales, so as not to send the "wrong message" to Taiwan's authorities at a sensitive time.[48] The next month, Ma greeted former U.S. National Security Adviser Brent Scowcroft and emphasized the good momentum between the two sides, saying that China appreciated the U.S. side's clear, open statement on its position on a referendum on United Nations membership for Taiwan.[49]

Ma also has developed a reputation for what are sometimes termed hawkish views on regional security issues on China's periphery, but this perspective appears to represent the mainstream view of the disputes in the Chinese security policymaking community. At the Shangri-La Dialogue in 2010, Ma declared that China would not accept surveillance aircraft in the East or South China Seas and that U.S. arms sales to Taiwan were "creating obstacles" in the Sino–U.S. military-to-military relationship. He also called for an end to Cold War–era alliances—a common Chinese reference to U.S. bilateral relations with not just Asian countries but countries around the world.[50] Amid a 2012 dispute with Philippine fishermen near Scarborough Shoal, Ma told a Hong Kong television station that "the South China Sea issue is none of the United States' business; it's just [territorial] disputes between China and its

---

[47] With the reorganization of MRs into "War Zones" or "Theater Area Commands" announced in late 2015, it is possible that the previous restriction on PLAAF officers being promoted to MR command will not carry over into the new "War Zones" or "Theater Area Commands." As a result, the assessment made here that DCGS assignment is required for top-tier PLAAF advancement may change.

[48] "China-US Talks," *China Daily*, December 6, 2007.

[49] Li Donghang [李东航], "Ma Xiaotian Meets US Guests" [马晓天会见美国客人], *Liberation Army Daily* [解放军报], January 22, 2008.

[50] Chan, 2012.

neighboring countries."[51] Since assuming command of the PLAAF, however, Ma does not appear to have traveled overseas at all in any capacity. He is widely cited as meeting with his counterparts from abroad while participating in PLA meetings in Beijing and elsewhere in China, but either Ma is hesitant or unwilling to travel or the leadership organs have not seen fit to have him represent PLAAF overseas in his new capacity. If one believes that the PLAAF could benefit from senior officer exposure to foreign air forces, then Ma's reluctance to travel, or bureaucratic restrictions on his ability to do so, may be a limiting factor on these military-to-military and service-to-service relations.

Xu and Ma's Potential Legacy on the PLAAF: An Expanded Set of Strategic Missions

Xu's Legacy: Unprecedented Responsibility and Ventures into the Space Domain

Xu Qiliang's perch on the CMC may well dictate the PLAAF's budgetary and force development success or failure for decades, but precedent for Xu is scant. The only other non-PLAA officer elevated to vice chairman in recent decades was General Liu Huaqing, who served from 1989 until his retirement in 1997 following his service as PLAN commander; at that time, he replaced his PLAN uniform with an Army uniform for his CMC service. Technically, Liu was not even a "career" PLAN officer, having entered service in the army—the Peasants' Red Army—in 1930, some 19 years before the formation of the PLA. He subsequently took part in the Long March as part of the 25th Red Army in 1935 and solidified his revolutionary credentials as commander of the unit that escorted future leader Deng Xiaoping from the Eighth Route Army Headquarters to the 129th Division to assume a new post during the War of Resistance Against Japanese Aggression.[52]

Most importantly for this discussion, Liu is regarded as the officer most responsible for setting the PLAN on a course of steady, extended modernization of maritime strategy, equipment, and tactics, and he is specifically remembered as the father of China's aircraft carrier. While he began in the Army, Liu's steady rise up the PLAN's ranks started in 1952, and he held a steady succession of acquisition-related and other senior Navy positions, starting in the mid-1960s and culminating in his rise to Navy commander in August 1982. At this point, his official biography credits Liu with two key accomplishments: First, Liu led a rectification campaign for the Navy that "brought a dramatic change to the Navy's image," and second, he oversaw the Navy's armament plan for the seventh Five-Year Plan (1986–1990), a publication titled "Navy in 2000," and the Program for Developing Naval Armaments.[53] Taken together, these efforts and programs essentially positioned the PLAN for the 21st century. In short, the tasks facing Liu as

[51] Chan, 2012.

[52] Xinhua, "Comrade Liu Huaqing's Biography" [刘华清同志生平], January 24, 2011.

[53] Xinhua, 2011.

he rose to the top of PLAN sound very much like those on the PLAAF's current task list. And although it is difficult to say with certainty, it seems reasonable that Ma will have the primary responsibility and authority for implementing and overseeing PLAAF programs on the CMC until the end of his term in 2017, while Xu is positioned to champion the service's cause until 2022.

Due in part to the strategic decisions taken some 20 years ago by the CMC regarding the importance of the PLAN and the extension of China's maritime capabilities, the PLAN was tapped to extend its operational space and capabilities, to include routinely completing antipiracy tasks in the Gulf of Aden; since 2008, the PLAN has deployed 20 antipiracy task groups to the region. Last year, the task group rotating on duty engaged in a red force versus blue force confrontational exercise designed in part for "researching and exploring difficult problems in distant sea operations training under informatized conditions."[54] The PLAN's steady presence in the gulf allowed it to complete an evacuation of over 500 Chinese citizens and a handful of foreign nationals employed by Chinese enterprises from Yemen in March 2015.[55] Days later, Chinese forces evacuated from Yemen 225 people from ten different countries, marking the first time the Chinese government helped with the evacuation of foreigners under deteriorating security conditions.[56]

There have been no sustained missions at great distances like the multiyear antipiracy duty for the PLAAF to this point, but the PLAAF has completed important support tasks that broaden its operational capabilities. For example, it has been active in the search for downed Malaysian Airlines flight MH-370 and also sent four Il-76 transports to help evacuate hundreds of Chinese citizens from Libya in March 2011.[57] But these missions predominantly have been simulated through exercises, which, while becoming increasingly realistic and reportedly involving more-complex tasks, are not the same as operating at these distances and stretching maintenance, sustainment, and command and control capabilities over these distances and time frames. Echoing the PLAN experience from 25 years ago, PLAAF leaders under Xu's command spent an inordinate amount of time in rectification campaigns designed to improve all facets of PLAAF force and personnel development, the most extensive of which was announced in 2008 and appears to have run through 2010.[58] (We discuss the PLAAF's duties as they relate to China's Air Defense Identification Zone [ADIZ] in the East China Sea in the next section of this paper.)

Meanwhile, real-world events and the PLAAF training plan for 2015 have the PLAAF stretching its operational tempo and responsibilities somewhat, and in meaningful ways. Burma's

---

[54] CCTV News [CCTV 新闻], August 27, 2014.

[55] "Chinese Naval Taskforce Completes Evacuating Chinese Nationals from Yemen," *China Military Online*, PRC Ministry of National Defense, March 31, 2015.

[56] Xinhua, "China Helps 10 Countries Evacuate Nationals from Yemen," April 3, 2015b.

[57] "Military Aircraft Bring Back 287 Chinese from Libya," *People's Daily Online,* March 5, 2011.

[58] "The Air Force Held New Outline Training Session and Combined Tactical Training Symposium in Beijing, Commander Xu Qiliang Attended and Gave a Speech," *Beijing Kongjun Bao*, October 28, 2008.

inadvertent bombing of Chinese territory in March 2015 afforded regional PLAAF units some real-world operations experience, as air defense, early warning units, and combat aircraft deployed to the border area by mid-March 2015.[59] Of more long-term significance for PLAAF and U.S. forces operating in the Western Pacific in peacetime, the PLAAF for the first time trained in the western Pacific in late March, when it sent an undetermined number of H-6 aircraft to the Bashi Channel, between southern Taiwan and the northern Philippine islands.[60] Subsequent reports from Chinese military officials stated that such drills would become standard practice and that the PLAAF and PLAN would "join together" for such exercises in the future.[61]

Yet Xu is most likely to be remembered for shepherding the PLAAF's operations into the domain of space and for its expanded mission set. It was Xu who first publicly announced China's plans to develop weapons in space on the occasion of the 60th anniversary of the PLAAF in November 2009, saying, "China's national interests are expanding and the country has entered the age of space. The (Chinese Communist) Party and the people have given us a historic mission. After thorough consideration, we decided to change." Xu also took the opportunity to roll out the tenets of the PLAAF's strategic goals and its use of space capabilities: "The Air Force will extend its reach from the sky to space, from defense of Chinese territory to attack (of threats) as well. We will improve the overall capability to strike a long-distance target with high precision, fight electronic or Internet warfare with back-up from space . . . and deliver our military strategic assets."[62] Five years later, Ma used the PLAAF's anniversary to rephrase these goals slightly to reflect Xi Jinping's priorities, describing the goal as a "powerful people's Air Force for integrated air and space operations that is capable of attack and defense and of *providing a strong support for the realization of the China dream and the dream of making the armed forces strong* [emphasis added]."[63]

The space domain is not currently one in which the PLAAF is the lead service, but at least some in the Air Force have argued for a consolidation of air and space missions under PLAAF, and specifically in the book *Strategic Air Force*. While not an official Professional Military Education textbook, this volume makes the case that air forces in many nations are responsible for space operations and that China's other services are insufficient fits for the mission because

---

[59] Jeffrey Lin and P. W. Singer, "China Mobilizes Forces on Burmese Border: Response to Fatal Bombing by Burma," *Popular Science*, March 17, 2015.

[60] "PLA Air Force Conducts First Training in West Pacific," *China Military Online*, March 30, 2015.

[61] Liang Jun, ed., "PLA's Air Force Drill over the High Seas to Be a Standard Practice," *People's Daily*, April 2, 2015.

[62] Stephen Chen and Greg Torode, "China to Put Weapons in Space," *South China Morning Post*, November 3, 2009.

[63] Ma Xiaotian [马晓天] and Tian Xiusi [田修思], "Speed Up the Building of a Powerful People's Air Force for Integrated Air and Space Operations That Is Capable of Attack and Defense—Studying Chairman Xi Jinping's Important Expositions on the Building and Development of the Air Force" [加快建设一支空天一体攻防兼备的强大人民空军—学习贯彻习近平主席关于空军建设发展的重要论述], *Seeking Truth* [求是], October 30, 2014, emphasis added.

their systems and platforms do not operate in the aerospace realm.[64] There is little evidence that this argument is gaining traction in any public debate, however; China's space operations remain the purview of PLA writ large, rather than any one service, and remain so under the sweeping military reforms announced in early 2016.[65] But other points from *Strategic Air Force* are in evidence in PLAAF and joint training since its publication in 2009. These include increased training for regional use of PLAAF and joint forces in exercises and especially the ubiquity of PLAAF (and PLA) reliance on information systems in the "system-of-systems" confrontations it expects to fight in the future.

Ma's Legacy: Maritime Security Moves to the Fore, While Strategic Goals Endure

Ma Xiaotian's comments in April 2014 were also clear about the PLAAF's expanded mission set, which preceded Xi but has gained immediacy under his leadership. In addition to echoing the PLAAF's need to accelerate its timeline for developing offensive and defensive capabilities, Ma emphasized maritime rights and the maritime domain, an operational space that has traditionally not been under the PLAAF's purview. Ma wrote that the Air Force must

> be fully and clearly aware that the air actions for safeguarding *maritime rights and interests* bring about higher requirements for guidance on the employment of the Air Force as well as Air Force force building and comprehensive support; be fully and clearly aware of the importance of *winning the initiative in air struggle for effectively coping with various security threats from the maritime domain*; be fully and clearly aware that *the new situation in the maritime rights* defense struggle brings about *new requirements for the Air Force to quicken its transformation* from homeland air defense to possessing both offensive and defensive capabilities [emphasis added].[66]

Ma's preoccupation with the maritime domain acknowledges that perhaps the most important near-term security developments for the PLAAF have been the renewed tensions over the Senkaku (Diaoyu) Islands since 2012 and China's subsequent declaration of an ADIZ in the East China Sea, which the PLAAF has been responsible for patrolling since China formally announced its creation in November 2013. (Thus far, operational responsibility for the South China Seas has been led by the PLAN since China began dramatically expanding its claims there in 2014.) The new ADIZ gave the PLAAF, for the first time, an operational patrol space well away from China's borders; according to a PLAAF spokesman, the Air Force completed 333 patrols in the ADIZ in 2013[67] and reportedly began flying airborne early warning and control

---

[64] Zhu Hui, ed., *Strategic Air Force* [战略空军论], Beijing, China: Blue Sky Press, 2009.

[65] The military reforms include the formation of a Strategic Support Force (SSF), which appears to have some of the responsibilities for space issues that had formerly been the purview of the General Armaments Department. Other space issues were subsumed under the CMC. The SSF is responsible for space, cyber, and electronic warfare. For a fuller discussion of the SSF, see John Costello, "The Strategic Support Force: China's Information Warfare Service," *China Brief*, Vol.16, No. 3, 2010, from the Jamestown Foundation.

[66] Ma Xiaotian, 2014, p. 6.

[67] Zhu Ningzhu, ed., "Chinese Military Planes Regularly Patrol ADIZ: Spokesman," Xinhua, January 23, 2014.

patrols in January 2014.[68] Meanwhile, Japan's Defense Ministry reports that Japanese aircraft scrambled to respond to Chinese aircraft operating in the area 464 times in fiscal year 2014.[69] With little fanfare, the PLAAF has been adding to its technical capabilities and responsibilities for air space management and early warning for the past decade, including taking charge of a joint air surveillance system over the East China Sea in 2006 and adding new air defense radars by 2010.[70]

The PLAAF is adapting, albeit slowly, to the new operational demands, as many represent a sharp departure from any mission tasks it had been held accountable for in the past. With no recent history of extensive overwater operations beyond its air surveillance and air warning roles, the PLAAF lacks the experience, specialized equipment, and training for maritime operations. Maritime search and rescue, for example, has never been an Air Force mission area. At least in part as a result of the PLAAF's lack of experience in this area, Ma found himself in a very unusual training setting in a November 2014 visit to Guangzhou MR Air Force units: on a ship at an unidentified maritime training base. Ma, as head of a PLAAF Working Group, reviewed the "first Air Force maritime unit" at its new training base and was briefed on PLAAF maritime search and rescue exercises (see Figure 2).[71] He emphasized the importance of command efficiency and the clarification of command coordination relationships in joint military-civilian search and rescue operations—all new territory for the PLAAF. Ma's comments also stressed building search and rescue aircraft, developing new systems for search and rescue operations, and ensuring the compatibility and integration of existing civil and military systems.[72]

---

[68] Yang Lina, ed., "Airborne Early Warning Aircraft to Serve over East China Sea," Xinhua, January 27, 2014.

[69] "Japan Scrambles Fighters 943 Times in FY 2014, 2nd Highest on Record," *Kyodo News*, April 15, 2015. Japan's fiscal year runs from April 1 to March 30.

[70] Mark Stokes, "China's ADIZ System: Goals and Challenges," Thinking-Taiwan.com, April 24, 2014.

[71] "While Inspecting a Certain Guangzhou Military Region Air Force Unit, Ma Xiaotian, Member of the Central Military Commission and Commander of the Air Force, Emphasized Joint Maritime Search and Rescue Will Require Integration of Military and Local Organizations and High Command Efficiency" [中央军委委员、空军司令员马晓天在广空部队检查调研时强调海上联合搜求要军地一体多元融合指挥高效], *Air Force News* [空军报], November 10, 2014, p. 1.

[72] "While Inspecting a Certain Guangzhou Military Region Air Force Unit, Ma Xiaotian, Member of the Central Military Commission and Commander of the Air Force, Emphasized Joint Maritime Search and Rescue Will Require Integration of Military and Local Organizations and High Command Efficiency," 2014.

successfully direct a complicated informatized joint operation, and it is necessary to rely on an integrated command information system."[80]

For the PLAAF, then, the emphasis on integrated joint operations has meant not just learning how to use complex information systems for command and control, but learning how to work and fight seamlessly in new domains with other services. As previously mentioned, these new domains explicitly include space and the PLAAF's use of information technology and space-based systems to meet its operational goals. It does not, however, mean that the PLAAF has been given broad new responsibilities for control of the space domain, but rather that PLAAF operational capabilities have become more capable with and reliant on these new data and information flows enabled by China's dramatic investments in space. How well—and how quickly—the PLAAF achieves its objectives in integrated joint operations will shape the legacies of both Xu and Ma and determine whether Xi's "strong army" dream will be realized.

Conclusion: Will New Leadership Equal Strategic Success for the PLAAF?

Xu Qiliang and Ma Xiaotian represent the new image of the PLAAF: They share broad operational experience, including in Xu's case what passes for combat experience; possess China's version of seasoning in joint staff positions; and, given that they have survived and thrived under Xi Jinping to this point, have been judged either sufficiently "red" or well-connected enough to remain in key leadership positions. Indeed, the preoccupation with the anticorruption drive could eventually conflict with China's ambitious goals for revamping training, which show no signs of abating even as general officers (and many others in the ranks) are taken into custody. Moreover, some command approaches, designed to educate joint commanders currently being taught at PLA postgraduate institutions, emphasize risk taking and seizing much more initiative than the PLA has typically allowed; it would not be unreasonable to think that some career officers would in fact be *less* willing to do so in such an environment.

But the success of Xu's and Ma's leadership tenures will be determined more by the PLAAF's progress toward achieving ambitious strategic goals than by the way they weather the current political environment. These goals stretch into the middle of this century, and judging from their comments as covered by official Chinese media outlets, PLAAF leaders understand the task before them. The PLAAF is now in the second stage of a three-stage development process; between 2010 and 2030, the PLAAF expects to develop some capacity for "certain air and space attack and defense capabilities," according to this plan.[81] Extending into the space domain, whether PLAAF does indeed receive a larger portion of that mission bureaucratically, will require a long-term commitment to new operational approaches and a training regimen to support these approaches. Official press accounts reflect that there is no short cut or quick fix

---

[80] Wang Wowen, 2010.

[81] Zhu Hui, 2009.

and that cultural change—"work style" in PLA parlance—is called for, but these same shortfalls have been systemic and well known for years. In fact, Xu and Ma have been grappling with this cultural change as PLAAF leaders in earnest since a new training outline was announced in 2008.[82]

It remains to be seen if Xu represents—accurately or not—the type of charismatic, transformative leader for the PLAAF that Liu Huaqing was for the PLAN. On the one hand, he appears to be an operationally capable officer and an adept leader—not to mention a survivor in a treacherous political system. On the other hand, the bribery and graft charges leveled against Xu Caihou suggest that all current senior officers were almost certainly aware of and arguably complicit in the corruption in the system, given that they navigated through it in some way. Moreover, Xu falls short of Liu's clout within the CCP, as the latter remains the last military officer to hold the most senior CCP positions, having served as a member of the Politburo and Politburo Standing Committee from 1992 to 1996.[83]

While Xu's background does not have the same resonance as Liu's bona fides and ties to leadership dating to Deng Xiaoping and the Long March, his promotion to vice chairman of the CMC—unprecedented for a member of the PLAAF—and appointment as chair of the CMC discipline inspection leading group at such a tumultuous time appear to demonstrate the trust Xi and other senior Party leaders have in Xu. While Ma may be the face of PLAAF modernization for now, Xu is uniquely positioned to promote and protect PLAAF interests and equities, though to date there is no public evidence that Xu has used his role as vice chairman as a platform to advocate on the PLAAF's behalf. Leaving public promotion of PLAAF to the service commander could simply reflect the duties and Xu's and Ma's respective responsibilities, and we have no way of knowing for certain what transpires at CMC meetings. But Xu's appointment probably signals that the PLAAF has gained a degree of status and influence, and his stature in the military hierarchy bodes well for the PLAAF in the bureaucratic competition for resources, potentially enabling it to make progress toward achieving its enhanced mission set over the next decade.

---

[82] Liu Pengyue [刘鹏越], "The Air Force Held a New Outline Training Session and Combined Tactical Training Symposium in Beijing" [空军新大纲集训暨合同战术训练研讨会在京举行], *Air Force News* [空军报], October 28, 2008, p. 1.

[83] Kenneth Allen, "Assessing the PLA's Promotion Ladder to CMC Member Based on Grades Vs. Ranks—Part 2," *China Brief*, The Jamestown Foundation, Vol. 10, No. 15, August 5, 2010b.

References

"The Air Force Held New Outline Training Session and Combined Tactical Training Symposium in Beijing, Commander Xu Qiliang Attended and Gave a Speech," *Beijing Kongjun Bao*, October 28, 2008.

Allen, Kenneth, "Assessing the PLA's Promotion Ladder to CMC Member Based on Grades vs. Ranks—Part 1," *China Brief*, Vol. 10, No. 15, July 22, 2010a. As of March 28, 2016: http://www.jamestown.org/single/?tx_ttnews%5Btt_news%5D=36660&no_cache=1

———, "Assessing the PLA's Promotion Ladder to CMC Member Based on Grades Vs. Ranks—Part 2," *China Brief*, Vol. 10, No. 15, August 5, 2010b. As of June 15, 2016: http://www.jamestown.org/programs/chinabrief/single/?tx_ttnews%5Btt_news%5D=36720&cHash=a9e4d0099f#.V2GzD1eFGao

Allen, Kenneth, Dennis J. Blasko, and John F. Corbett, "The PLA's New Organizational Structure: What is Known, Unknown and Speculation (Part 1)," *China Brief*, Vol. 16, No. 3, February 4, 2016. As of August 10, 2016: http://www.jamestown.org/single/?tx_ttnews%5Btt_news%5D=45069&no_cache=1#.V6mpvT5962w

Allen, Kenneth W., Glenn Krumel, and Jonathan D. Pollack, *China's Air Force Enters the 21st Century*, Santa Monica, Calif.: RAND Corporation, MR-580-AF, 1995. As of July 13, 2016: http://www.rand.org/pubs/monograph_reports/MR580.html

Bishop, Bill, "A Highly Public Trip for China's President, and Its First Lady," *New York Times*, March 25, 2013. As of May 4, 2015: http://dealbook.nytimes.com/2013/03/25/china-dream-apple-nightmare

CCTV News [CCTV 新闻], August 27, 2014.

Chan, Minnie, "PLA Hawk Ma Xiaotian Flying Ever Higher," *South China Morning Post*, October 25, 2012.

Chen, Stephen, and Greg Torode, "China to Put Weapons in Space," *South China Morning Post*, November 3, 2009.

Cheng Li, "China's Midterm Jockeying: Gearing Up for 2012 (Part 3: Military Leaders)," *China Leadership Monitor*, No. 33, 2010. As of July 13, 2016: http://www.hoover.org/sites/default/files/uploads/documents/CLM33CL.pdf

Chin Chien-li [金千里], "One of the Key People in Chinese Communist Combat Operations Against Taiwan: Critical Biography of the General Staff Lieutenant General Xu Qiliang" [中共对台作战中坚人物：中共副总参谋长许其亮中将评传], *The Frontier* [前线], June 1, 2005.

———, "The CPC's Key Figure in Conflict with Taiwan: A Critical Biography of Lieutenant General Zhang Qinsheng, the New Deputy Chief of General Staff—Prefect of Hu Jintao's 'Class of Military Generals'" [中共对台作战中坚人物：新任副总参谋长章沁生中将评传—胡锦涛的'将军版'班长], *The Frontline* [前线], March 1, 2007, pp. 78–84.

"China-US Talks," *China Daily*, December 6, 2007.

"Chinese Naval Taskforce Completes Evacuating Chinese Nationals from Yemen," *China Military Online*, PRC Ministry of National Defense, March 31, 2015. As of May 5, 2015: http://eng.mod.gov.cn/TopNews/2015-03/31/content_4577869_3.htm.

Choi Chi-yuk, "General Fan Changlong Tipped for Top Post in China Military Commission," *South China Morning Post*, October 22, 2012a.

———, "PLA Leadership Changes Reflect a Fine Balancing Act," *South China Morning Post*, October 26, 2012b.

Costello, John, "The Strategic Support Force: China's Information Warfare Service," *China Brief*, Vol. 16, No. 3, 2010.

Guo, Aibing, "China's Top Anti-Graft Regulator to Inspect 26 State Companies," *Bloomberg Business*, February 11, 2015. As of May 4, 2015: http://www.bloomberg.com/news/articles/2015-02-12/china-s-top-anti-graft-regulator-to-inspect-26-state-companies

"Handle the Trial of Gu Junshan's Case Strictly on the Basis of Law—PLA Expert Yu Xiao Interprets Relevant Legal Issues in the Trial of Gu Junshan's Case" [严格依法审理谷俊山案件—军队法律专家欲晓解读审理谷俊山案件有关诉讼法律问题], *Liberation Army Daily* [解放军报], April 1, 2014.

"In Depth Study and Implementation of Chairman Xi Jinping's Important Expositions on National defense and Army Building, Pushing Forward the Great Practice of Building and Strengthening the Military From a New Starting Point" [深入学习贯彻习主席关于国防和军队建设重要论述，在新的起点上推进强军兴军伟大实践], *Liberation Army Daily* [解放军报], April 2, 2014.

Information Office of the State Council, "China's National Defense in 2006," December 2006. As of May 5, 2015: http://www.china.org.cn/english/features/book/194421.htm

"Japan Scrambles Fighters 943 times in FY 2014, 2nd Highest on Record," *Kyodo News*, April 15, 2015.

Kor Kian Beng, "18 PLA Generals Show Collective Support for China's President Xi," *The Straits Times*, April 3, 2014.

Lescot, Patrick, *Before Mao: The Untold Story of Li Lisan and the Creation of Communist China*, trans. Steven Rendall, New York: HarperCollins Publishers, 2004.

Levin, Dan, "China Names 14 Generals Suspected of Corruption," *New York Times*, March 2, 2015. As of March 31, 2015: http://www.nytimes.com/2015/03/03/world/china-names-14-generals-suspected-of-corruption.html

Li Donghang [李东航], "Ma Xiaotian Meets US Guests" [马晓天会见美国客人], *Liberation Army Daily* [解放军报], January 22, 2008.

Liang Jun, ed., "PLA's Air Force Drill over the High Seas to Be a Standard Practice," *People's Daily*, April 2, 2015. As of May 5, 2015: http://en.people.cn/n/2015/0402/c98649-8872999.html

Lin, Jeffrey, and P. W. Singer, "China Mobilizes Forces on Burmese Border: Response to Fatal Bombing by Burma," *Popular Science*, March 17, 2015. As of May 5, 2015: http://www.popsci.com/china-mobilizes-forces-burmese-border

Liu Pengyue [刘鹏越], "The Air Force Held a New Outline Training Session and Combined Tactical Training Symposium in Beijing" [空军新大纲集训暨合同战术训练研讨会在京举行], *Air Force News* [空军报], October 28, 2008.

Ma Cheng-kun, "Assessing the New PLA Leadership," presentation at CAPS-RAND-NDU Conference on the PLA, November 2014.

Ma Hao-Liang [马浩亮], "New Records in the Promotion of Central Military Commission Vice Chairmen" [军委副主席普生创纪录], *Ta Kung Pao* [大公报], November 5, 2012.

Ma Xiaotian [马晓天], "Strive to Enhance the Air Force's Ability to Fight and Win" [努力提高空军部队能打仗打胜仗能力], *Liberation Army Daily* [解放军报], April 2, 2014.

Ma Xiaotian [马晓天] and Tian Xiusi [田修思], "Speed Up the Building of a Powerful People's Air Force for Integrated Air and Space Operations That Is Capable of Attack and Defense—Studying Chairman Xi Jinping's Important Expositions on the Building and Development of the Air Force" [加快建设一支空天一体攻防兼备的强大人民空军—学习贯彻习近平主席关于空军建设发展的重要论述], *Seeking Truth* [求是], October 30, 2014.

Mao Zedong, "On Correcting Mistaken Ideas in the Party," December 1929.

Meng, Angela, "PLA Reshuffle Strengthens Xi Jinping's Hand in Corruption Fight," *South China Morning Post*, September 22, 2014. As of May 5, 2015: http://www.scmp.com/news/china/article/1597643/pla-reshuffle-strengthens-xi-jinpings-hand-corruption-fight

"Military Aircraft Bring Back 287 Chinese from Libya," *People's Daily Online,* March 5, 2011.

Mulvenon, James, "Hotel Gutian: We Haven't Had That Spirit Here Since 1929," *China Leadership Monitor*, No. 46, Winter 2015. As of March 31, 2015: http://www.hoover.org/sites/default/files/research/docs/clm46jm.pdf

Osnos, Evan, "Born Red," *The New Yorker*, April 6, 2015. As of May 4, 2015: http://www.newyorker.com/magazine/2015/04/06/born-red

Oster, Shai, "President Xi's Anti-Corruption Campaign Biggest Since Mao," *Bloomberg Business*, March 4, 2014. As of May 28, 2015: http://www.bloomberg.com/news/articles/2014-03-03/china-s-xi-broadens-graft-crackdown-to-boost-influence

"'Party Commands Gun' Must Be Upheld," *Liberation Army Daily*, November 3, 2014. As of March 31, 2015: http://english.chinamil.com.cn/news-channels/china-military-news/2014-11/03/content_6209449.htm

"Personnel Changes at the People's Liberation Army General Logistics Department: Gu Junshan Is No Longer Deputy Director" [解放军总后勤部人事调整，谷俊山不再担任副部长], *People's Daily* [人民日报], February 11, 2012.

Pew Research Center, "Environmental Concerns on the Rise in China: Many Also Worried About Inflation, Inequality, Corruption," September 19, 2013. As of May 4, 2015: http://www.pewglobal.org/files/2013/09/Pew-Global-Attitudes-Project-China-Report-FINAL-9-19-132.pdf

"PLA Air Force Conducts First Training in West Pacific," *China Military Online*, March 30, 2015. As of May 5, 2015: http://english.chinamil.com.cn/news-channels/2015-03/30/content_6420538.htm

PLA General Political Department [中国人民解放军总政治部], "The Scientific Guide to Strengthening and Developing the Military Under the New Situation—Deeply Study and Implement Chairman Xi Jinping's Important Speech at the All-PLA Political Work Conference" [新形势下强军兴军的科学指南—深入学习贯彻习近平主席在全军政治工作会议上的重要讲话], *Seeking Truth* [求是], December 1, 2014. As of March 31, 2015: http://www.qstheory.cn/dukan/qs/2014-12/01/c_1113408713.htm

"PLA Inspection Tour Organs Established, PRC Ministry of National Defense," *China Military Online*, November 22, 2013.

"PLA Senior Generals Back Xi's Orders," *China Daily*, April 3, 2014. As of May 4, 2015: http://www.chinadaily.com.cn/china/2014-04/03/content_17404124.htm

Reuters, "China Investigates Three Senior Army Officers Suspected of Graft," April 27, 2015. As of May 5, 2015:
http://uk.reuters.com/article/2015/04/27/uk-china-corruption-idUKKBN0NI07520150427

———, "China's Xi Sets Up Five New 'Battle Zones' in Military Reform Push," February 1, 2016. As of March 28, 2016:
http://www.reuters.com/article/us-china-military-reform-idUSKCN0VA2EO

"Scores of PLA Officers Punished," *China Daily*, January 30, 2015.

Stokes, Mark, "China's ADIZ System: Goals and Challenges," Thinking-Taiwan.com, April 24, 2014. As of June 1, 2015:
http://thinking-taiwan.com/chinas-adiz/

Wang Wowen [王握文], "Information Capability: Primary Capability of Joint Operations—Interview with Kuang Xinghua, Professor and Doctoral Adviser at the National Defense Science and Technology University" [信息能力：联合作战的第一能力—访国防科学技术大学教授、博士生导师匡兴华], *Liberation Army Daily* [解放军报], May 27, 2010.

"While Inspecting a Certain Guangzhou Military Region Air Force Unit, Ma Xiaotian, Member of the Central Military Commission and Commander of the Air Force, Emphasized Joint Maritime Search and Rescue Will Require Integration of Military and Local Organizations and High Command Efficiency" [中央军委委员、空军司令员马晓天在广空部队检查调研时强调海上联合搜求要军地一体多元融合指挥高效], *Air Force News* [空军报], November 10, 2014.

Wong, Edward, "Xi Jinping's News Alert: Chinese Media Must Serve the Party," *New York Times*, February 22, 2016. As of March 24, 2016:
http://www.nytimes.com/2016/02/23/world/asia/china-media-policy-xi-jinping.html

Xie Shenzong [解审综] and Sai Zongbao [塞宗宝], "Pushing Forward Innovative Development of Armed Forces' Audit Work from New Starting Point—Roundup on All Army and Armed Police Units Focusing on Strong-Military Objective To Strengthen Audit Work" [在新起点上推动军队审计工作创新发展—全军和武警部队聚集强军目标加强审计工作综述], *Liberation Army Daily* [解放军报], January 29, 2015.

Xinhua, "China's PLA Eyes Future in Space, Air: Air Force Commander," November 1, 2009.

———, "Comrade Liu Huaqing's Biography" [刘华清同志生平], January 24, 2011.

———, "Fan Changlong Conducts Fact-Finding at Tibet- and Qinghai-Based Army Units, Emphasizes Need to Deeply, Solidly Conduct Live Combat Military Training, Constantly Increase Army Units' Ability to Perform Their Functions and Missions" [范长龙在驻西藏、

青海部队调研时强调深入扎实抓好实战化军事训练，
不断提高部队履行职能任务能力], August 17, 2014a.

———, "All Army Political Work Conference Begins in Gutian; Xi Jinping Attends and Gives Important Speech" [全军政治工作会议在谷田召开，习近平出席会议并发表重要讲话], November 1, 2014b.

———, "Xu Caihou Dies of Bladder Cancer," March 15, 2015a.

———, "China Helps 10 Countries Evacuate Nationals from Yemen," April 3, 2015b.

———, "President Xi Visits China's National News Outlets," February 19, 2016.

Yang Lina, ed., "Airborne Early Warning Aircraft to Serve over East China Sea," Xinhua, January 27, 2014. As of May 5, 2015:
http://news.xinhuanet.com/english/video/2014-01/27/c_133077734.htm

Yang Ting [杨婷], ed., "With the Approval of Xi Jinping, the Central Military Commission Prints and Issues the 'Decision of the Central Military Commission on Unfolding Inspection Work' and the 'Regulations of the Central Military Commission on Inspection Work (Trial'); Xu Qiliang Chairs a Meeting to Make Implementation Arrangements" [经习近平主席批准中央军委引发《中央军委关于开展巡视工作的决定》《中央军委巡视工作规定（试行），许其亮主持召开会议对贯彻落实作出部署], Xinhua [新华], October 29, 2013.

Yuan Banggen, "Setting Eyes on Development, Stepping Up Research in Information Warfare Theories and Construction of Digital Force and Digital Battlefields," *China Military Science* [中国军事科学], February 20, 1999, pp. 46–51.

Zhang Hong, "PLA Generals Take Rare Step of Swearing Loyalty to President Xi Jinping," *South China Morning Post Online*, April 3, 2014. As of May 4, 2015:
http://www.scmp.com/news/china/article/1463386/pla-generals-take-rare-step-swearing-loyalty-president-xi-jinping

Zhang Yuqing [张玉清] and Li Xuanliang [李宣良], "Xi Jinping Inspects the PLA Air Force Organs, Stresses More Quickly Building a Powerful People's Air Force Characterized by Air and Space Integration, Both Offense and Defensive Capabilities So as to Provide Strong Force Support for the Fulfillment of the Chinese Dream and the Strong Army Dream" [习近平在空军机关调研时强调加快建设一支空天一体攻防兼备的强大人民空军为实现中国梦想强军梦提供坚强力量支撑], Xinhua [新华], April 14, 2014.

Zhao Yinan, "Xi Repeats Anti-Graft Message to Top Leaders," *China Daily*, November 20, 2012. As of May 4, 2015:
http://usa.chinadaily.com.cn/china/2012-11/20/content_15942824.htm

Zhu Hui, ed., *Strategic Air Force* [战略空军论], Beijing, China: Blue Sky Press, 2009.

Zhu Ningzhu, ed., "Chinese Military Planes Regularly Patrol ADIZ: Spokesman," Xinhua, January 23, 2014. As of May 5, 2015: http://news.xinhuanet.com/english/china/2014-01/23/c_133069151.htm

Chapter Two

Trends in PLA Air Force Joint Training

Assessing Progress in Integrated Joint Operations

Mark R. Cozad and Astrid Stuth Cevallos

Summary

In light of the People's Liberation Army's (PLA's) limited combat experience in recent years, Chinese military analysts have paid close attention to the form and outcomes of U.S. joint operations in the former Yugoslavia, Iraq, and Afghanistan. The lessons they learned from these conflicts led to a series of new operational concepts to remedy shortfalls in areas in which the PLA Air Force (PLAAF) either was relying on antiquated practices or had no experience altogether, including integrated command and control; intelligence, surveillance, and reconnaissance (ISR); and information and electromagnetic warfare. Several of the core operational concepts that emerged from these observations have focused on the PLAAF's participation in joint multiservice exercises and its ability to perform critical missions, such as precision strikes, intelligence gathering, command and control, and air defense. Accordingly, the PLAAF's development of its capabilities and its role in implementing these operational concepts are core components of the PLA's evolving framework for "integrated joint operations" (一体化联合作战). As a result, PLAAF leaders at all levels have increasingly stressed more-realistic exercises, greater pilot autonomy and flexibility, improved logistics support, and the ability to discover and attack enemy targets in a contested and uncertain environment.

Joint exercises—like the ones examined in this report—have given the PLA, and PLAAF leaders more specifically, opportunities to test progress toward achieving these objectives, leaving them with an improved understanding of their shortcomings and of the need for practical experience in planning and executing more-realistic training. It is unclear, however, whether subsequent training events have implemented these lessons in any meaningful way. Although more-recent exercises suggest some level of improvement and increased sophistication, it also remains unclear just how sophisticated PLAAF capabilities have become. Certainly, the systemic shortfalls identified in several PLA training events—difficulty in coordination, obstacles to information sharing, limited realism, and a continued reliance on scripting—remain persistent problems for PLA and PLAAF leaders, hampering key functions such as air defense, targeting, battle damage assessment, and air support to ground forces. Absent significant changes that provide more-realistic and more-complex training, the PLAAF's capabilities will continue to be constrained by its outdated organizational culture.

Defining Joint Operations: Lessons Learned and Key Concepts

Introduction: Learning About Joint Operations from Others

The 1990s were a watershed in the history of the PLAAF. Military and civilian leaders in the People's Republic of China (PRC) closely observed U.S. operations in the Middle East and North Atlantic Treaty Organization (NATO) operations in southeastern Europe and realized how far their forces had fallen behind these technologically advanced militaries. Concerned that China's forces were unprepared for modern combat, PRC analysts studied the successes and failures of the Gulf War and Kosovo campaigns, drawing lessons for the PLA about "asymmetrical war" and "local wars under high tech conditions," focusing particularly on air strike and counter–air strike operations.[84] Their research led to a wholesale restructuring of all PLA services that encompassed a new military strategy, new operational concepts, the pursuit of advanced technologies, and accelerated purchases of advanced Russian weapons and platforms.[85]

In particular, the operational surprises and resulting lessons learned from U.S. air operations in the Gulf War and Kosovo compelled the PLAAF to confront its weaknesses and step up modernization efforts. Since the PLAAF's inception, the PRC's air defense strategy had been centered on short-range point air defenses with limited night capability and mobility that were focused on protecting critical targets, generally key military and industrial facilities.[86] But this strategy rapidly collapsed as U.S. operations demonstrated that modern air forces equipped with precision-guided weapons; advanced command, control, communications, computers, intelligence, surveillance, and reconnaissance (C4ISR); and aerial refueling offered an unprecedented threat to the survivability of the PRC's most strategically significant infrastructure.[87]

Many PLA observers thus argued that offensive air operations would define future conflicts, requiring the PLAAF to invest in precision munitions and ISR to enhance its air defense

---

[84] Two key studies include Huang Bin, *Research into the Kosovo War,* Beijing: Liberation Army Publishing House, 2000; and Wang Yongming, Liu Xiaoli, and Xiao Yunhua, *Research into the Iraq War,* Beijing: Military Science Publishing House, 2003. These two studies, along with numerous other military science journal and military press publications, highlight the PLA's intense focus on foreign military developments, particularly those involving the United States. As this paper will discuss, many of these lessons are captured in PLA operational concepts developed to respond to specific operational requirements.

[85] See David Finkelstein, "China's National Military Strategy: An Overview of the 'Military Strategic Guidelines,'" in Roy Kamphausen and Andrew Scobell, eds., *Right-Sizing the People's Liberation Army: Exploring the Contours of China's Military*, Carlisle, Pa.: Institute for Strategic Studies, May 2007, for a discussion of the Military Strategic Guidelines and the central role they play in delineating planning and modernization requirements. See also James C. Mulvenon and Andrew N. D. Yang, eds., *A Poverty of Riches: New Challenges and Opportunities in PLA Research*, Santa Monica, Calif.: RAND Corporation, CF-189-NSRD, 2004.

[86] Ge Dongsheng, ed., *On National Security Strategy*, Beijing: Military Science Publishing House, 2006, pp. 231–233.

[87] Peng Guangqian and Yao Youzhi, eds., *The Science of Military Strategy,* Beijing: Military Science Publishing House, 2005, pp. 321–322; and Ge Dongsheng, 2006, p. 234.

posture.[88] They noted that U.S. and allied aircraft had easily gained and maintained the initiative in each conflict, deploying with impunity around the periphery of the countries under attack and operating with little concern for defending their own assets against attack.[89] In the face of such overwhelming adversary air power, the PLA could no longer rely on massive numbers of ground forces. Instead, future conflicts would depend on the significant maneuverability and destructive capacity that air power provided. In the 1993 revision of the *Military Strategic Guidelines in the New Era* (still in force today, though with subsequent modifications), these observations coalesced into the core objective of conducting integrated joint operations, a concept that has since guided the development of new systems and operational concepts.[90]

This paper examines the factors and concepts driving PLAAF joint training and evaluates the PLAAF's progress toward achieving this critical component of its long-term objectives. In particular, it analyzes the development of joint operational concepts, subsequent experimentation with these new ideas, and their integration into major exercises and training events. This research is based primarily on open-source Chinese military texts—including military science publications and military newspapers—that detail the broad range of PLA training activities designed to develop and operationalize these emerging concepts.

Developing New Operational Concepts Based on U.S. Experience

In light of the PLA's limited combat experience in recent years, the PLA has paid close attention to the form and outcomes of U.S. joint operations in Iraq, the former Yugoslavia, and Afghanistan. These and other examples have underscored the importance of integrated command and control, ISR, and information and electromagnetic warfare.[91] Numerous publications have documented these lessons in detail, often tying them to the broad contours of PLA modernization programs and foreign procurement efforts.[92] Instead of providing another overarching discussion, this section examines the manner in which these lessons were tied into operational concepts. Taken together, these examples provide a framework for understanding the PLA and PLAAF process for developing new operational concepts, testing those concepts, and then promulgating findings across the force in the form of accepted tactics and training practices. Each of the following concepts were directly related to the PLA's observations from U.S. operations and illustrate the PLA and PLAAF emphasis on the importance of jointness.

---

[88] Wang Houqing and Zhang Xingye, eds., *The Science of Campaigns*, Beijing: National Defense University, 2001, pp. 418–422; Ge Dongsheng, 2006, pp. 231–235.

[89] Huang Bin, 2000, pp. 140–144; and Wang Yongming, Liu Xiaoli, and Xiao Yunhua, 2003, p. 199. In addition, see Ge Dongsheng, 2006, p. 234; and Zhang Yuliang [张玉良], ed., *The Science of Campaigns* [战役学], Beijing: National Defense University Press [国防大学出版社], 2006, p. 97.

[90] Wang Houqing and Zhang Xingye, 2001, p. 63.

[91] Hao Yuqing and Cai Renzhao, eds., *Science of Armed Forces Building*, Beijing: National Defense University Press, 2007, pp. 280–282, and Ge Dongsheng, 2006, pp. 234–236.

[92] See Huang Bin, 2000; and Wang Yongming, Liu Xiaoli, and Xiao Yunhua, 2003.

Three Attacks, Three Defenses

The notion of "three attacks, three defenses" (三打三防) formed the conceptual foundation for most PLA training in the aftermath of U.S. operations in the former Yugoslavia.[93] Formulated in response to Operation Allied Force, the concept spawned a number of training events and an expansive discussion about the importance of "scientific training."[94] "Three attacks, three defenses" called for attacking with precision-guided munitions, cruise missiles, and attack helicopters while defending against electronic warfare, stealth aircraft, and ISR. Although many of the insights that originally shaped the concept were flawed or poorly understood and initial training events were questionable in their effectiveness, the concept was refined over time and eventually provided solid insights into the types of air threats against which the PLAAF would have to defend in any future conflict with the United States.

Joint Antiair Raid Campaign

Closely related to the idea of "three attacks, three defenses," the "joint antiair raid campaign" (联合反空袭战役) concept was devised to enhance national air defense.[95] Indeed, the primary purpose for this campaign concept was to defend China's national airspace and vital targets against the advanced capabilities that challenged the PLAAF's existing inventory and force posture. From the outset, this concept was recognized as necessarily joint, requiring capabilities from across the PLA for both offensive and defensive operations that would hinder an adversary's ability to conduct strikes against strategic targets in the PRC.[96] The PLAAF viewed the development of both offensive and defensive capabilities as essential to avoiding the problems that had confronted Serbia and Iraq while under U.S. military attack.[97] The long-held notion that PLAAF assets could remain tied to their bases near key targets was no longer feasible. PLAAF munitions, sensors, C4ISR, and training were just a few of the many areas in which PLAAF leaders were forced to confront the systemic weaknesses of an obsolete, legacy

[93] See Fu Quanyou, "Deepening Military Training by Scientific and Technological Means," Xinhua, June 27, 2000. This editorial from the then–PLA Chief of the General Staff highlighted the importance of the "three attacks, three defenses" in the PLA's overall training program. Most notably, he stated that the PLA needed to give high priority to training that effectively addressed the technological needs associated with modern warfare. This editorial was published immediately following an All-Army Conference that examined the integration of science and technology into training.

[94] Ren Chaohai [任朝海], "Accomplish the Transformation of Theoretical Results into Work Results" [实现理论成果向工作成果转化], *Liberation Army Daily* [解放军报], November 22, 2000; Chen Youyuan [陈有元], "Deepen Training in the New 'Three Attacks and Three Defenses'" [深化新'三打三防'训练], *Liberation Army Daily* [解放军报], November 14, 2000; and Xu Sheng [徐升], "Panorama Talk on Science and Technology Military Exercises" [科技练兵纵横谈], *Liberation Army Daily* [解放军报], August 10, 1999.

[95] Wang Houqing and Zhang Xingye, 2006, pp. 331–334; and Li Yongsheng, *Lectures on the Science of Joint Campaigns*, Beijing: Military Science Press, 2012, p. 242. The antiair raid campaign is defined in the *Lectures on the Science of Joint Campaigns* as one of the main campaign forms for defensive campaigns.

[96] Ge Dongsheng, 2006, pp. 233–234.

[97] Ge Dongsheng, 2006, p. 231.

force. The joint antiair raid campaign thus provided an organizational and conceptual framework for providing an extended defensive capability to confront an enemy's advanced systems farther from China's mainland while simultaneously employing offensive weapons against enemy airfields and logistics facilities to threaten force build ups.

Noncontact Warfare

"Three attacks, three defenses" and the joint antiair raid campaign concepts were reflections of a more sweeping development signified by U.S. operations—the emergence of "noncontact warfare" (非接触作战). According to several senior PLA writers who helped develop the concept, this new form of warfare constituted a significant departure from earlier models of warfare in several important respects.[98] Until the 1990s, they argued, warfare was based on a model of attrition that sought the destruction of fielded forces; military success was primarily achieved by mass deployments of mechanized forces. U.S. operations in the former Yugoslavia demonstrated that modern, high technology–enabled warfare no longer conformed to this model. The objective of military operations had changed from attrition to the destruction of an enemy's war potential, embodied in strategic targets like leadership, energy, industry, communications, and key infrastructure.[99] Long-range precision strikes on these targets, enabled by advanced C4ISR capabilities, would be the cornerstone of modern warfare. The noncontact warfare model required PLA commanders to bring together each service's firepower capabilities in unprecedented ways. PRC leaders and PLA commanders also had to consider the entire range of kinetic and nonkinetic capabilities at their disposal. Airpower played a central role in the development of noncontact warfare. Furthermore, the ability to employ airpower over long ranges with precision weapons, advanced command and control, and ISR made it an indispensable component of future capabilities.

Target-Centric Warfare

The most recent PLA operational concept under development is "target-centric warfare" (目标中心战), which appears to be a further refinement of the noncontact warfare model. The general concept behind target-centric warfare is that by employing ISR sensors and target analysis, PLA commanders can identify—and subsequently aim to destroy—the most critical targets in a system.[100] This operational concept seeks to make efficient use of firepower assets, provide timely targeting of the most essential targets on the battlefield, and ensure that combat plans are able to adapt in an agile manner that addresses rapid changes in a dynamic

---

[98] See Liu Yuejun, ed., *Non-Contact Warfare,* Beijing: Military Science Publishing House, 2004; Pan Youmu, *The Study of Non-Contact Warfare,* Beijing: National Defense University Publishing House, 2003; and Shao Guopei, ed., *Information Operations in Non-Contact Wars,* Beijing: Liberation Army Publishing House, 2004.
[99] Liu Yuejun, 2004, pp. 8–9; and Pan Youmu, 2003, p. 50.
[100] Liu Shenyang [刘沈扬], "The Theory and Practice of Target-Centric Warfare" [目标中心战的理论与实践], *China Military Science* [中国军事科学], No. 5, 2013.

environment. Recent target-centric warfare experimentation has focused on engaging mobile targets and employing opposition forces to challenge exercise participants.[101] As this concept is still under development, there has been only a limited amount of literature available describing its evolution and key elements.

Defining Integrated Joint Operations: The Role of Informatization and System-of-Systems Operations

These operational concepts arising from PLA observations of foreign military operations have focused on the PLAAF's participation in joint multiservice operations and its ability to perform critical missions, such as precision strikes, intelligence gathering, command and control, and air defense. Accordingly, the PLAAF's development of its capabilities and its role in implementing these operational concepts are core components of the PLA's evolving framework for "integrated joint operations" (一体化联合作战), the concept that forms the foundation of PLA thinking on joint operations. The PLA textbook *Science of Campaigns* defines *integrated joint operations* as "using integrated methods and information technology, blending an operational system from all services and arms and other types of armed strengths with operational units to form an integrated whole."[102] Integrated joint operations take a variety of forms in multiple battlespace dimensions with the ultimate objective of seizing control of the "three superiorities," controlling the battlefield, and destroying the enemy's operational system.[103] Accordingly, integrated joint operations rely on a flexible system that permits and enables adjustments and coordination over the entire depth and within every battlespace domain as the situation requires. As one senior PLA officer argued, these types of operations are driven by "the guiding ideology of 'comprehensive supremacy, precision strike, and destruction of systems.'"[104]

As the PLA's concept of integrated joint operations emerged, it was tied to and embedded in two other key ideas that drive PLA modernization objectives—"informatization" (信息化) and "system-of-systems operations" (体系作战). To understand how the PLAAF is building its

---

[101] Ding Yahan [丁雅涵] and Li Dezhong [李德忠], "In This Battle, Aiming at the 'Vulnerable Spot' When Firing – Close-Up View of the Military Region's 'Queshan Decisive Victory-2013A' Exercises with Troops, Part III" [这一丈，瞄准"七寸"打 － 近观军区"确山决胜－2013A"实兵演习之三], *Vanguard News* [前卫报], December 8, 2013; and Chang Xin [常欣], "Use Military Innovation Theory to Guide Substantial Leap in Combat Power" [以军事创新理论牵引战斗力实质跃升], *Vanguard News* [前卫报], March 31, 2013.

[102] Wang Houqing and Zhang Xingye, 2006, p. 80.

[103] Wang Houqing and Zhang Xingye, 2006. The "three superiorities" described in PLA literature include information, air, and sea superiority. This concept is an essential component in PLA campaign design and a critical precondition for ensuring operational success.

[104] Zhan Yu [战玉], "A Study of the Theory of Integrated Joint Operations" [一体化联合作战理论探要], *China Military Science* [中国军事科学], No. 6, 2007, pp. 11–21.

critical role in integrated joint operations, this paper examines the PLAAF's participation in joint exercises with respect to these two additional critical ideas.

The concept of informatization has guided PLA modernization for at least the past decade.[105] The PLA considers it to be the essence of integrated joint operations, which rely on information networks to integrate and systematize operations designed to obtain information superiority.[106] From this perspective, informatization underpins most facets of integrated joint operations and serves as a key unifying theme in much of the experimentation supporting the development of new operational concepts, including "three attacks, three defenses," noncontact warfare, and target-centric warfare. Integrated joint operations are considered "the basic form and necessary requirement for informatized war."[107] In particular, the idea of informatization recognizes the indispensable role that information and information systems will play in future wars. Success in these operations will ultimately depend on ensuring that information on incoming attacks, emerging targets, force disposition, and intentions is available for all parts of the joint force. Advances in Western technology—particularly in stealth, precision, and long-range employment—further require that critical information be passed in a timely fashion over great distances. These requirements form the basis for how PLA exercises over the past decade have sought to bring together critical capabilities from different services and branches within the PLA to develop operational concepts. In this respect, informatization in these exercises has centered on information dominance achieved through robust, common C4ISR systems, information sharing, and information protection. In a recent visit to a PLAAF organization, Central Military Commission (CMC) Vice Chairman Fan Changlong stated that

> It is necessary to firmly establish the idea of information dominance. Given the fact that there is more and more informatized equipment, commanding organs at all levels and all officers and men must strengthen the concept of information dominance, strive to learn knowledge about informatization, master informatized equipment, and improve skills related to informatization.[108]

Accordingly, informatization will continue to serve as the basis for much of the PLAAF's joint training and will be an important guiding concept tied to the PLA's idea of system-of-systems. The PLAAF's emphasis on the "system-of-systems" concept is based on ensuring joint capability by building critical links in command automation, ISR, precision strike, and rapid mobility, which have been shown in recent military operations to be the backbone of modern

---

[105] Ge Dongsheng, 2006, p. 62; Hao Yuqing and Cai Renzhao, p. 280; and Wang Houqing and Zhang Xingye, 2006, p. 85.

[106] Song Youfa and Hong Yaobin, eds., *Integrated Joint Operations Command Headquarters Work,* Beijing: Military Science Press, 2005, p. 1.

[107] Wang Houqing and Zhang Xingye, 2006, p. 80.

[108] Xiong Weiwei, "Seriously Implement the Series of Important Decisions and Instructions of Chairman Xi and Making Persistent Efforts in Deeply Pushing Forward Actual Combat-Oriented Training" [认真贯彻习主席系列重要决策指示再接再厉深入推进实战化训练], *Air Force News* [空军报], October 27, 2014.

warfare.[109] It seeks to employ "combat systems" that integrate those capabilities and systems necessary for a specific operational purpose, usually defined in the context of joint or combined campaigns. More specifically, *The Campaign Theory Study Guide*, a key PLA textbook, identified the connection between campaigns and combat systems in the following manner (emphasis added):

> Paralyzing the enemy's combat system has become an important means of winning a war . . . . Once there are problems in key links of the system, the entire weapon system and combat system will lose its combat effectiveness, or will even become paralyzed. This illustrates that modern campaigns are the confrontation between combat systems. Advanced weapons and equipment and good strategy and planning both depend upon the integrity and coordination of combat systems. Therefore, in modern campaigns, *attacking and paralyzing key nodes in the enemy's combat system* while ensuring the integrity and coordination of one's own combat systems has become an important way of winning.[110]

As this important text points out, building and protecting one's own combat system while simultaneously identifying the critical weaknesses of an adversary's systems and attacking them are the two imperatives for success in future military operations. Accordingly, PLAAF training and exercises have used this central concept as a guide to building the capabilities necessary to achieve the PLA's most important operational concepts.

Testing Concepts: Major Joint Exercises and Training Events

This section highlights PLAAF participation in several major joint exercises in which the PLA has attempted to operationalize the core concepts previously discussed. It is not a comprehensive analysis of all joint exercises over the past ten years, but rather an overview of mostly recent exercises that illustrates how key ideas were tested in the PLA's experimentation process and then implemented within and across military regions (MRs). In some cases, data regarding the specific units involved in these exercises and the details of their activities are lacking. However, as the PLA has built these concepts—and its overarching concept for integrated joint operations—it has published exercise descriptions and after-action reports in military newspapers to disseminate key findings and ideas across the PLA, aiming to popularize these practices with PLA commanders, as well as with the rank-and-file.

In each case, the PLAAF has been instrumental in the development of these new operational concepts. As part of this process, PLAAF leaders at all levels have placed increasing emphasis on more-realistic exercises, greater pilot autonomy and flexibility, improved logistics support, and the ability to discover and attack enemy targets in a contested and uncertain environment.

---

[109] Xiong Weiwei, 2014.

[110] Xue Xinglin [薛兴林], ed., *Campaign Theory Study Guide* [战役理论学习指南], Beijing: National Defense University Press [国防大学出版社], 2001, p. 66.

Although these themes do not always emerge in the details of reports carried in the PLA press, they are generally discussed as essential elements of the operational concepts as they develop. In most cases, the content represents some attempt on the PLAAF's part to enrich and improve its training. That said, it is often difficult to get adequate detail to evaluate how much progress PLAAF units have actually made.

Since 2005 and the first Sharp Sword exercise exploring integrated joint operations, PLAAF capabilities have been a core component in several of the PLA's joint exercises. Due to the number of exercises during this time, this report explores a few exercises that contributed to the development of many of the operational concepts outlined above, particularly those in which the PLAAF has played a prominent role. These exercises span multiple MRs and demonstrate that many of the leading efforts in concept development take place in out-of-the-way locations. The following sections briefly describe the PLAAF's participation in joint exercises in six of the seven MRs—Beijing, Chengdu, Guangzhou, Jinan, Lanzhou, and Nanjing—to provide a glimpse into how PLAAF modernization is supporting these emerging ideas.

Early Experimentation on Integrated Joint Operations, 2004–2005

In 2004, the PLA General Staff Department designated the Chengdu MR as one of the test centers for experimentation on integrated joint operations.[111] Numerous research topics were assigned as part of the exploratory project, including combat capabilities, training, and integrated joint actions. A *Liberation Army Daily* article recounted the exercise's specific objectives as follows:

> In accordance with the notions of "integrated interaction between fighting, building, and training, and integrated joint actions of services and arms," the Chengdu MR created training contents beneficial to generating systemic capabilities, built systems platforms to enable sharing of information between services and arms, explored methods and means of integrated joint training, and set up cross-services and arms and cross-establishments joint training mechanisms.[112]

To support the project, the MR's leaders designated lead units from both the PLA Army and PLAAF to participate. Experts from across the PLA, most of whom were military science researchers and academics, provided additional support. Shijiazhuang Army Command College partnered with the MR in capturing and developing the theoretical guidelines and practices as they emerged from the experimentation.[113]

---

[111] Cheng Sixun, "Exploration and Practice of Integrated Training of Military Region Units: Part One," *Battle Flag News* [战旗报], February 9, 2006a.

[112] Zhuang Lijun [庄利军], "A Rapidly Expanding Transformation in the Training Domain" [训练领域一场方兴未艾的变革], *Liberation Army Daily* [解放军报], February 6, 2006.

[113] Cheng Sixun, "Exploration and Practice of Integrated Training of Military Region Units: Part Three," *Battle Flag News* [战旗报], February 14, 2006c.

The experiment's overall intent was to improve the joint operations capabilities of the ground and air forces in the MR through the development of a joint command and control system and to build and integrate "emergency mobile combat units."[114] More specifically, a major focus for the experiment was to develop an integrated combat system that linked dispersed combat units from different services into an integrated whole. During this process, the MR developed a series of documents that outline general guidelines that included an "Overall Plan of Experimental Integrated Training," "Opinions on Strengthening the Building of Combat Capabilities of Units in the Military Region Under Informatized Conditions," a "Development Program for the Comprehensive Integration of Emergency Mobile Combat Units," and a plan for "Constructing a System of Integrated Training and Training Methods Based on Research and Exchange Activities of All-Army Integrated Training Experimentation."[115] The overall experimentation process culminated in the Sharp Sword exercise and demonstration held in October 2005.

A parallel experimentation effort in integrated joint operations also ran at roughly the same time in the Nanjing MR.[116] This effort involved elements of PLAA units in the Nanjing MR, as well as PLAAF and Second Artillery units.[117] Its main purpose was to experiment with "internal integration within services and arms to strike a foundation, coordinated training of services and arms in techniques, joint training of services and arms to promote fusion, and joint training of services and arms to strengthen integration."[118]

Both exercises demonstrated the process by which the PLA sought to build its integrated joint operations efforts around specific operational scenarios and the PLAAF's multiple supporting roles.

Cross-MR Mobility and the Mission Action-Series Exercises

China's 2013 Defense White Paper, entitled "The Diversified Employment of China's Armed Forces," described cross–MR exercises (跨区机动演习) as one of the three key categories of "scenario-based" exercises that the PLA conducts to "develop rapid response and joint

---

[114] Cheng Sixun, "Exploration and Practice of Integrated Training of Military Region Units: Part Two," *Battle Flag News* [战旗报], February 10, 2006b.

[115] Cheng Sixun, 2006b.

[116] Zhuang Lijun, 2006; Zhang Wenping and Yan Wenbo, "Advance Phase of Second Artillery's Integrated Training Starts—Establishing Steering Group on Integrated Training, Organizing Trial Comprehensive Integration and Integrated Training, Conducting Theoretical Study on Integrated Combat and Training, and Exploring Characteristics and Laws of Integrated Training," *Rocket Forces News* [火箭兵报], July 13, 2004; and Lu Feng and Ni Menzhi,"Mobile and Camouflaged Launches Using New Equipment Under Complex Weather and Terrain Conditions," *People's Front* [人民前线], July 28, 2004.

[117] As a result of the PLA's recent reorganization, the PLA's Second Artillery Force is now known as the PLA Rocket Force. All subsequent references to Second Artillery are based on the organization's name at the time of the event.

[118] Zhuang Lijun, 2006.

operations capabilities in unfamiliar environments and [under] complex conditions."[119] The document then explicitly named the campaign-level Mission Action (使命行动) exercises, conducted annually since 2010, as examples of such long-distance mobility training. While the groundbreaking Mission Action-2010 ground-and-air joint exercise conducted by group armies and PLAAF units of the Beijing, Chengdu, and Lanzhou MRs received extensive coverage in official media outlets, two brief, vague sentences on the 2011 and 2012 exercises (the latter of which also involved the PLAAF) in the Defense White Paper appear to be the only official acknowledgment of these exercises' occurrence. The widespread reporting on the Mission Action-2010 and the more recent Mission Action-2013 exercises not only provides a wealth of information about the PLAAF's ability to conduct cross-MR integrated joint operations, but also suggests that PRC leaders might regard these exercises as important benchmarks of PLAAF progress.

This section compares the Mission Action exercises that took place in 2010 and 2013. Both of these cross-MR joint campaign exercises emphasized unscripted, actual force-on-force combat scenarios and joint attacks in complex electromagnetic environments. Perhaps most important, these exercises were critical tests of the PLAAF's ability to conduct integrated joint operations with PLAA and even PLAN forces (in the case of Mission Action-2013) using an information technology–based integrated command platform (一体化指挥平台).

"'Jointness' Was Everywhere": Mission Action-2010

The landmark Mission Action-2010 exercise marked the debut of this integrated command platform in large-scale campaign-level joint combat operations.[120] Centrally planned and organized by the four PLA general departments as an opportunity for units to explore and practice transitioning from regional defense to whole territory combat maneuvers,[121] Mission Action-2010 was the final large-scale military exercise of the 11th Five-Year Program (2006–2010).[122] As such, it was designed to evaluate group army commanders' progress in conducting

---

[119] Information Office of the PRC State Council [中华人民共和国国务院新闻办公室], "Defense White Paper: The Diversified Employment of China's Armed Forces" [国防白皮书：中国武装力量的多样化运用], April 16, 2013.

[120] Liang Hongqing [梁洪清], "Integrated Command Platform Employed in Large-Scale Real Combat for the First Time" [一体化指挥平台首次大规模投入实战应用], *Battle Flag News* [战旗报], October 21, 2010. The integrated combat platform is a software suite that provides joint command automation tools and functions for both legacy and newly deployed C4 systems.

[121] Yang Hong [杨鸿], "Military Region Conducts Exercise 'Mission Action-2010A' Oath Swearing and Mobilization Rally" [军区举行'使命行动－2010A'演习誓师动员大会], *Comrades-in-Arms News* [战友报], October 12, 2010.

[122] Liu Fengan [刘逢安] and Wu Tianmin [武天敏], "Training Transformation: Striving to Advance Amid Wind and Rain—Interview with Chen Zhaohai, Director of Military Training and Service Arms Department of PLA General Staff Department" [训练转变：在风雨中奋力前行—访总参军训和兵种部部长陈照海], *Liberation Army Daily* [解放军报], November 2, 2010.

multiservice “system-of-systems operations” (体系作战)[123] according to the objectives laid out in the January 2009 edition of the *Outline of Military Training and Evaluation* (军事训练预考核大纲), including improving proficiency in confrontational training (对抗训练) and training in complex electromagnetic environments (复杂电磁环境下训练).[124]

As Table 1 shows, the ground and air joint exercise was divided into three phases, each commanded by a different MR’s group army (军团) and each organized as a force-on-force confrontational exercise. Each MR air force (MRAF) participated throughout all phases, providing transport and air support for ground forces and carrying out airborne reconnaissance, electronic jamming, and air-to-ground firepower strikes. Roughly 30,000 troops participated across all three phases.[125]

Table 1. Three Phases of Mission Action-2010

Title	Dates	Number of Participants	Primary/Leading Participant	Objective
2010A	October 10–21	10,000	Beijing MR 27th Group Army	Move from Beijing MR Zhurihe Combined Arms Tactical Training Base (CATTB) to Shenyang MR Taonan CATTB (~1,500 km); practice joint campaign command, ground-air long-distance movement, and joint firepower attacks in complex electromagnetic environments, unfamiliar terrain, and changing weather[a]
2010B	October 18–28	10,000	Lanzhou MR 47th Group Army	Move from Lanzhou MR Qingtongxia CATTB to Chengdu MR Xichang CATTB (~2,000 km); defend against network attack
2010C	October 12–21	10,000	Chengdu MR 13th Group Army	Move from Chengdu MR Xichang CATTB to Lanzhou MR Qingtongxia base (~2,000 km)[b]

[a] Yang Hong, 2010.
[b] Zhang Li and Zhu Qianli, “Air Force Units Fulfill Stage Tasks in the Trans-Region Mobile Exercise; Precision Command Drives Ground-Air Joint Operations,” *Liberation Army Daily*, October 27, 2010.

While both Mission Action-2010A and Mission Action-2010C received extensive coverage in MR newspapers and national news media, Mission Action-2010B was less widely covered. An article in the Lanzhou MR newspaper, *People’s Army* (人民军队), highlighted one unique aspect of Mission Action-2010B: The exercise, directed by a group made up of PLA general department staff, featured a specially created “‘Blue Force’ network warfare detachment” that probed the Red Force command network for vulnerabilities, eventually determining that the

[123] Liu Fengan and Wu Tianmin, 2010.

[124] Liu Fengan [刘逢安] and Wu Tianmin [武天敏], “New-Generation ‘Outline of Military Training and Evaluation’ Issued” [新一代《军事训练预考核大纲》颁发], *Liberation Army Daily* [解放军报], July 24, 2008.

[125] “Live News” [新闻直播], CCTV Xinwen, October 9, 2010.

network was "equipped with tight security measures, and effective reconnaissance, penetration, and further attack were impossible."[126]

The coverage of the exercises in the Beijing and Chengdu MRs showed broad similarities; the main thread was that—as one newspaper noted—"'Jointness' was everywhere."[127] Two overarching themes dominated the media coverage of the Mission Action-2010A and Mission Action-2010C exercises. First, reporters and their interlocutors underscored the importance of joint command and control using the informatized, integrated command platform. A CCTV report on the air force's participation in all Mission Action-2010 exercises specified that joint campaign command organizations in each of the MRs contained command teams from the participating MRAFs.[128] These multiservice command structures were aided in their efforts to guide joint operations involving ground and air units by the integrated command platform, which enabled them to share intelligence in real time and bypass middlemen in communicating orders directly to the forces.[129] An unnamed staff officer from the Chengdu MRAF Headquarters Department's Operations Division described the innovation's importance for synchronizing joint operations in these terms:

> In the past, command and control was like stacking toy building blocks (搭积木): Each level had to report the situation and transmit orders down the chain of command level by level. Now these matters travel through the integrated command platform directly to their final destination, making multilevel joint action and long-distance synchronization a reality.[130]

Gao Jianguo, the executive director of the Mission Action-2010A exercise and one of the deputy chiefs of staff of the Beijing MR, further described how operation orders from the joint campaign command organizations were issued to the MRAF (joint campaign large formation, 联合战役军团) and the air division (joint tactical formation, 联合战术兵团) through the integrated command platform, then issued by the air division commanders to the air force air units, army aviation units, signal units, scout units, engineer units, reconnaissance units, artillery units, tank units, armored units, and special operations forces.[131]

---

[126] Qing Yong and Mo Fuchun, "First Phase of Trans-Region Mobility Exercise 'Mission Action-2010B' Completed, Command Information System Subjected to Network Attack Throughout Entire Course," *People's Army* [人民军队], October 20, 2010.

[127] Han Guozhang [韩国章], "Inspiration of and Reflection on 'Mission-Action-2010A' Exercise" ['使命行动－2010A'演习的启示与思考], *Comrades-in-Arms News* [战友报], November 2, 2010.

[128] "Military Report" [军事报道], CCTV-7, October 16, 2010.

[129] Yang Qingmin, "'Mission Action-2010A' Trans-Region Mobile Exercise of the Beijing Military Region Shows Three Major Highlighting Features" [使命行动 2010A 北京军区跨区机动演习凸显三大亮点], Xinhua [新华], October 21, 2010.

[130] Liang Hongquing, 2010.

[131] Yang Qingmin, 2010. See also Kenneth W. Allen, "Introduction to the PLA's Administrative and Operational Structure," in James C. Mulvenon and Andrew N. D. Yang, eds., *The People's Liberation Army as Organization*, Santa Monica, Calif.: RAND Corporation, CF-182-NSRD, 2002.

Second, reports stressed that the exercises incorporated "actual combat" (实战), unfamiliar territory, and unexpected situations, including weather conditions and mishaps. Mu Guoxin, who was one of the deputy commanders of an air division of the Beijing MRAF, noted that Mission Action-2010A was "the first time for them to carry out the coordination between the air and the ground in an unfamiliar area and the first time for them to conduct a cross-MR long-distance maneuver exercise."[132] But it was not just the location and operations that were new: Deputy Regimental Commander Lu Hongbing added that, in its simulated struggle for command of the air, the air division "encountered the most complicated weather conditions in all of the joint exercises that they had ever participated in," as "rain and snow were nonstop and the clouds were low and thick."[133]

In each phase of the exercise, the PLAAF played a key role. Mu Guoxin remarked that the main PLAAF tasks during Mission-Action 2010 were providing "air cover, fighting for air dominance, and [conducting] air-to-ground assault."[134] Several different kinds of units were involved in a variety of tasks. In the Beijing MR, for example, the Blue Force consisted of reconnaissance aviation units that carried out aerial reconnaissance, surveillance, and simulated air strikes against the Red Force.[135] Electromagnetic warfare constituted another task; one report profiled a pilot named Ma Hongsheng, who "released an infrared jamming shell" in response to a "guided missile attack" from the enemy.[136] Mission Action-2010C reports were quite precise about the Chengdu MRAF's involvement in the exercise: Thirteen aircraft made 26 sorties over the course of eight days, carrying out air transport, aviation reconnaissance, electromagnetic interference, and air-to-ground attacks.[137]

As seen in the following section, the air force's involvement in Mission Action-2013 exercises expanded upon the foundation established during this earlier series of exercises.

"Information Secures Victory": Mission Action-2013

Like Mission Action-2010, the Mission Action-2013 exercise comprised several "firsts": It was the first cross-MR campaign exercise to involve amphibious landing operations and maritime force projection, and it was the first time that the PLAAF took on a leading role in a Mission Action exercise.[138] Unlike Mission Action-2010, in which two PLA scholars noted the

---

[132] Lin Hongguan, Yang Ping, and Liu Quan, "Assault in the Air and on the Ground—Account of the Participation of a Certain Aviation Division of the Beijing Air Force in the 'Mission Action-2010' Exercise," *Air Force News* [空军报], October 25, 2010.

[133] Lin Hongguan, Yang Ping, and Liu Quan, 2010.

[134] "Military Report," 2010.

[135] "Military Report," 2010.

[136] Lin Hongguan, Yang Ping, and Liu Quan, 2010, p. 1.

[137] Zhang Li and Zhu Qianli, 2010.

[138] Xinhua, "PLA Begins 'Mission Action-2013' Exercise on the Tenth" [解放军 10 日"使命行动－2013"演习], September 9, 2013.

ground forces were more or less doing a "one-man show" ("独角戏"), Mission Action-2013 saw a "change of the protagonist" (主角的变化) with another service—the PLAAF—in charge of part of the exercise, which increased overall jointness and integration.[139]

As Table 2 shows, the ground, air, and maritime joint exercise was divided into three phases, each commanded by a different MR's group army (军团). Each MRAF participated throughout all phases, providing transport and overseeing joint amphibious landings in coastal provinces; uniquely, the Guangzhou MRAF led the Mission Action 2013C phase, which included a long-range joint assault. Across the three phases, over 40,000 troops participated.[140]

Table 2. Three Phases of Mission Action-2013

Title	Dates	Number of Participants	Participants	Objective
2013A	September 10–19	17,000	Nanjing MR 31st Group Army, Nanjing MR Air Force, PLA Navy East Sea Fleet and South Sea Fleet, Second Artillery	Plan and execute long-range maneuvers on land and sea and in air; conduct joint amphibious landing exercise in Guangdong Province
2013B	October 11–20	20,000	Guangzhou MR 42nd Group Army, PLAN East Sea Fleet and South Sea Fleet, Guangzhou MR Air Force, 15th Airborne Corps	Respond to unexpected situations, including enemy jamming and bombing; conduct joint amphibious landing exercise in Fujian Province
2013C	September 21–27	10,000	Guangzhou MR Air Force	Deploy to airfields in coastal provinces, including Hainan; conduct long-range joint assault with almost 100 aircraft of different types

SOURCE: "Live News" [新闻直播], CCTV Xinwen, October 16, 2013a.

The Mission Action-2013 exercises improved upon many of the skills tested in the Mission Action-2010 exercises, including use of the integrated command platform, engaging in actual combat, and dealing with unexpected situations. Forecasting and dealing with inclement weather was a similar challenge: During Mission Action-2013A, a navy–air force joint weather forecasting tool helped predict the course of the Manyi and Usagi typhoons, providing the directors with information that allowed them to schedule the exercise for the lull in between the two typhoons.[141]

Long-distance maneuvers were once again an organizing theme, giving troops the chance to solve mobility-related "'bottleneck' issues" like "platform-free railway unloading [and] pontoon

---

[139] Kang Yongsheng [康永升] and Zhang Dianfu [张殿福], "The 'Mission Action' Military Exercise Has No 'Deep-Level Intentions'" ["使命行动"军演没有"深层用意"], *China Youth News* [中国青年报], September 27, 2013.

[140] "Live News," 2013a.

[141] Zhang Fenghai [张凤海] and Zhou Feng [周峰], "The Important Platform in the Military Strengthening Practice —Several Points of Revelation and Thought About the 'Mission Action-2013A' Cross-Region Mobile Campaign Exercise," *People's Front* [人民前线], October 18, 2013b.

setup for river-crossing."[142] An account of Mission Action-2013B alluded to another problem related to long-distance maneuvers for which the Chinese forces involved found a solution: Before this exercise, troop formations organized by unit and service traveled long distances separately, confronting any unexpected events with limited equipment and expertise. This time, however, the command information system, enabled "commanders of amphibious, Army Aviation, naval, and other units" to "break the boundaries between services" by creating "packaged" combined groups ("打包"编组) consisting of ground, naval, and air force troops.[143] The commanders split up more than 100 units and combined them together "like jigsaw puzzles" ("拼图形") to form combined groups with ideal combinations of specialized skills and equipment.[144]

Information, too, featured prominently throughout the exercise. One report on Mission Action-2013B referred to an "information cloud" ("信息云") over the battlefield that was created to strengthen the network command information system (网络指挥信息系统) throughout the area of the exercise.[145] In accordance with the Guangzhou MR's new notion of "information going out before the forces have moved" ("兵马未动，信息先行"), a hundred construction teams were sent out prior to the exercise to areas across four provinces and 12 cities, where they connected optical fiber circuits and set up additional 3G mobile stations and WiFi video systems.[146] These steps apparently increased access to and enhanced the quality of information received by the joint command, which included PLAN and PLAAF personnel.

Mission Action-2013C marked a significant development in PLA operations and thus deserves special attention. This phase of the exercise—led by the Guangzhou MRAF and observed by CMC member and PLAAF Commander Ma Xiaotian—is the only joint exercise directed by a PLAAF headquarters as of this writing.[147] While the other phases of Mission Action-2013 also incorporated air forces into three-dimensional (land, sea, and air) long-distance maneuvers and joint attacks with long-range artillery, army aviation, and amphibious tanks, Mission Action-2013C was more experimental in its approach, focusing more on joint air defense and confrontation in the air.[148] In one of the key components of the exercise, 93 PLAAF "Red Force" warplanes—early warning aircraft (including KJ-200s), refueling aircraft, fighter

---

[142] Zhang Fenghai and Zhou Feng, 2013b.

[143] Wang Wendui [王文锥] and Le Hongwang [骆宏望], "Troop Projection in 'Packages,' No 'Going Separate Ways'" ["打包投送"拒绝"各走各路"], *Soldiers News* [战士报], October 16, 2013.

[144] Wang Wendui and Le Hongwang, 2013.

[145] Mou Dong [牟东], Zhou Shiai [周世艾], and Liu Chang [刘畅], "For the First Time, the Military Region Proposes and Actually Constructs and Uses a Shaped Network Command Information System" [军区首次提出并实际构建运用栅格化网络指挥信息系统], *Soldiers News* [战士报], October 27, 2013.

[146] Mou Dong, Zhou Shiai, and Liu Chang, 2013.

[147] "Military Report" [军事报道], CCTV-7, September 22, 2013a.

[148] "Live News," 2013a.

jets (including J-10s), bombers, reconnaissance planes, and transport planes—took off from several different airfields in multiple provinces along the coast and, in collaboration with ground-to-air missile units, radar units, antiaircraft artillery, and paratroopers, traveled over long distances to reach the "Blue Force."[149] Early warning aircraft played a key role in the force-on-force confrontation that ensued between fighter aircraft: They carried out surveillance, coordinated with electronic jamming aircraft to suppress the Blue Force and provide an opening for the air combat formation's safe passage, and executed long-distance raids with dissimilar aircraft to penetrate the Blue Force's airspace.[150] These accomplishments inspired Red Force commander Xu Anxiang, who was then the Guangzhou MRAF commander, to describe the early warning aircraft as critical to incorporating, processing, and distributing information from all dimensions of the battlefield to "ensure unobstructed communications among the exercise's command organizations and a grasp of the troops' progress in real time," which he characterized as a "basic requirement of system-of-systems operations."[151]

More broadly, this component of the exercise was designed to realize the full potential of what a leader of the department directing the exercise called the "core operational capabilities of Air Force operations" (空军作战的核心能力)—long-distance raids (远程奔袭) and precision attacks (精确打击)—in combined and joint operations.[152] A passage from a CCTV "Military Report" (军事报道) broadcast about this long-distance joint assault operation illustrates how different aircraft of the Red Force used these capabilities to defeat the Blue Force in an intense battle at close quarters in maritime airspace:

> With the support of the early warning force and electronic warfare aircraft, the Red Force's fighter group shot down several Blue Force fighter planes. The Red Force's bomber formation successfully shook off interception by enemy aircraft and launched attacks against the Blue Force's ground targets. While the Red Force was crushing the Blue Force's low altitude penetration operation, its transport aircraft carrying the airborne operations detachment broke through the Blue Force's air defense and arrived in the airspace over the Blue Force's strategic point under the escort of several fighter planes. The paratroopers descended from the sky to attack and destroy the key targeted strongpoints.[153]

---

[149] "Live News," 2013a; "Military Report" [军事报道], CCTV-7, September 27, 2013b.

[150] "Military Report," 2013b; Zhang Li [张力], Wu Aili [武艾丽], and Zhao Lingyu [赵凌宇], "System-of-Systems Operation Directs Sword at Distant Sea—Witnessing the Air Force's Trans-Regional Maneuver Campaign Exercise 'Mission Action-2013C'" [提子作战剑指远海—空军"使命行动一2013C"跨区机动战役演习闻思录], *Air Force News* [空军报], November 1, 2013.

[151] Zhang Li, Wu Aili, and Zhao Lingyu, 2013.

[152] Zhang Li, Wu Aili, and Zhao Lingyu, 2013.

[153] "Military Report," 2013b.

A subsequent report notes that these paratroopers, apparently a special operations detachment of the PLA 15th Airborne Corps (under the jurisdiction of the PLAAF), successfully sabotaged the Blue Force's radar station and port shortly after landing.[154]

As in the Mission Action-2010 exercises—after which a mechanized infantry regiment commander stated that "speaking bluntly about problems, facing problems head-on, and solving problems are the only effective means to increase combat strength"[155]—relatively candid after-action conferences and their resulting reports documented shortcomings and errors during the Mission Action-2013 exercises. Even commanders, who were criticized for "talking about new operational concepts but using old ones" (作战理念讲的新用的旧) and "focusing on implementing decisions once they are made instead of adapting to changes" (定下决心后重执行轻变化), were not spared.[156] The command element and administrative and functional departments (指挥机关) struggled to use information systems to their full potential and to grasp something called the "four knows and one understand" ("四知、一个搞明白"), later explained as "know yourself, know the enemy, know the environment, know skills" ("知我、知敌、知环境、知技术") and "understand the mechanism for winning in combat under informatized conditions" ("搞明白信息化条件下作战制胜机理").[157] Units, too, were critiqued for their weak understanding of the enemy's situation, poor command of tactics, and inadequately organized combat plans. Referring specifically to the PLAAF, an *Air Force News* report noted that Mission Action-2013C exposed shortcomings in key areas of long-distance operations and joint interoperability.[158]

A brief, interesting note on this last point: While observers understandably might express concern about the increasing attention given to maritime joint operations in this exercise, academics from the PLA Academy of Military Science and the PLA Navy Marine Corps College cautioned foreign experts not to "over-interpret" an ulterior motive behind the location and nature of Mission Action-2013. They rejected the idea that this cross-MR maneuvering exercise was meant to act as a deterrent with regard to the Senkaku (Diaoyu) Islands and South China Sea disputes, instead framing the exercise as part of the PRC's "active defense" policy.[159]

---

[154] "Military Report" [军事报道], CCTV-7, November 25, 2013f.

[155] Han Guozhang, 2010.

[156] Zhang Fenghai [张凤海] and Zhou Feng [周峰], "Successful Conclusion of Trans-Regional Mobile Campaign Exercise 'Mission Action-2013A'" ["使命行动－2013A"跨区机动战役演习圆满落幕], *People's Front* [人民前线], September 20, 2013a.

[157] Zhang Fenghai and Zhao Feng, 2013a.

[158] Zhang Li, Wu Aili, and Zhao Lingyu, 2013.

[159] Kang Yongsheng and Zhang Dianfu, 2013.

Target-Centric Warfare and Nighttime Joint Operations Exercises in the Jinan Military Region

Ever since it was tasked to explore the new operational concept of "target-centric warfare" (目标中心战) in 2010, the Jinan MR has played a central role in developing and implementing the concept, which is closely linked to system-of-systems thinking. In October 2013, concurrent with the joint training and exercises described in this section, one of the Jinan MR deputy commanders, Lieutenant General Liu Shenyang, published an article in *China Military Science*, the journal of the PLA's Academy of Military Science, describing the "theory and practice" of the concept.[160] "In essence," Liu wrote, "target-centric warfare is a limited, quick war that relies on operational systems, is target-oriented, and involves striking key nodes."[161] He goes on to outline three specific ways in which this type of warfare is "target-centric": First, targets are selected for attack based on their function within an operational system; second, operational actions are oriented toward attacking those targets; and third, the progress of operations is determined based on the extent to which the intended target is damaged.[162] In practice, while targets may include the conventional casualties of combat—human lives and military assets—the concept focuses on the "attack, control, occupation or protection of the key information systems . . . used in combat," which may be destroyed by "firepower or deployment of forces" or by "electromagnetic or network assault."[163]

Table 3 shows two recent joint exercises conducted in the Jinan MR that explicitly focused on the concept of target-centric warfare. The first exercise, Queshan Decisive Victory-2013A (确山决胜-2013A), involved ground and air forces from the Jinan MR's 26th Group Army,[164] whereas the second and much larger exercise, Joint Decisive Victory-2013 (联合决胜-2013), involved nearly all forces in the Jinan MR theater, including the North Sea Fleet. While both exercises contained joint mountain offensive operations, Joint Decisive Victory-2013 focused far more on joint nighttime operations—both joint mountain offensive operations at night and tri-service joint beach seizing and landing at night—making it the PLA's first nighttime theater-scale joint operations exercise since theater joint training began in 2009.[165]

---

[160] Liu Shenyang, 2013, pp. 83–92.

[161] Liu Shenyang, 2013, p. 84.

[162] Liu Shenyang, 2013, p. 84.

[163] Song Linsheng [宋林胜], "Target Centric Warfare, a Deadly System-Breaking Weapon" [破击作战体系的利器 — 目标中心战], *China Military Online* [中国军网], February 6, 2013.

[164] Li Dezhong [李德忠] and Ma Yongsheng [马永生], "The 'Queshan Juesheng-2013A' Live Personnel Exercise Concludes Very Satisfactorily: Deepening Utilization of the Military Region Unit Training Reform and Innovation Achievements" ["确山决胜—2013A"实兵演习圆满落幕：深化运用军区部队训练改革创新成果], *Vanguard News* [前卫报], November 10, 2013.

[165] "Military Report" [军事报道], CCTV-7, November 30, 2013g; "Live News" [新闻直播], CCTV Xinwen, November 17, 2013c.

Table 3. Decisive Victory 2013 Joint Exercises in the Jinan Military Region

Title	Dates	Number of Participants	Participants	Objective
Queshan Decisive Victory-2013A	October 29–November 3	10,000	Jinan MR 26th Group Army; Jinan MR Air Force	Conduct ground and air joint tactical formation mountain offensive operations; conduct long-distance mobilization
Joint Decisive Victory-2013	November 16–19	18,000	Jinan MR Air Force; North Sea Fleet; Jinan MR 20th, 26th, and 54th Group Armies; Second Artillery	Conduct theater joint campaign army group tri-service nighttime operations, including a tri-service nighttime beach seizing and landing (first stage) and ground and air joint mountain offensive operations (second stage)

Facing a "Desperate Situation": Queshan Decisive Victory-2013A

In an article summing up Queshan Decisive Victory-2013A, Chen Zhaohai, one of the Jinan MR deputy commanders and an executive director of the exercise, cited the guiding role that target-centric warfare played in the exercises: Units chose strike targets based on their value to operational systems, focused on targets in organizing their operational actions, and adjusted their operations based on the destruction of targets.[166] Ding Feng, one of the deputy directors of the Jinan MR Headquarters Department's Military Training Department and an executive director of the exercise, echoed the importance of the concept and described the shift that it has effected in the focus and nature of operations, saying that "In the past, we placed importance on the scale of warfare and emphasized confrontation of massed forces. Now, we place more emphasis on elite troops seizing vital strategic points and destroying the opponent's system-of-systems."[167] To put these ideas about target-centric warfare into practice, operational command was conducted according to a four-stage process: "reconnaissance, control, strike, and evaluate" ("侦控打评").[168]

This process meant that reconnaissance and joint long-range precision strikes formed the core of the exercise. Reconnaissance satellites, air force reconnaissance aircraft,[169] and even unmanned aerial vehicles (无人机)[170] participated in the exercise—often on the side of the Blue Force, which tested the Red Force's concealment and camouflage, including the use of anti-

[166] Li Dezhong and Ma Yongsheng, 2013.

[167] Ding Yahan and Li Dezhong, 2013.

[168] Li Dezhong and Ma Yongsheng, 2013.

[169] Sun Jie [孙杰] and Fu Xiaohui [付晓辉], "Queshan Juesheng 2013A Exercise Enters Real-Troop, Live Fire Combat Stage" ["确山决胜-2013A"演习进入实兵实弹战斗阶段], *China National Radio* [中国广播网], November 4, 2013.

[170] Mei Changwei [梅常伟] and Li Dezhong [李德忠], "Exercise Starts This Time with a 'Scenario of Losing Battle'—Second Installment of Report Series on the Queshan Decisive Victory-2013A Exercises in the Jinan MR" [这一仗，从"残局"开打—近观军区"确山决胜－2013A－实兵演习之二], *Vanguard News* [前卫报], December 3, 2013.

infrared reconnaissance netting.[171] This strengthening of the opposing force appears to be the result of a new approach characterized by placing the Red Force in a "desperate situation" or "situation of losing battle" (残局) at the start of the exercise to truly test capabilities, move troops off script, and increase the realism of combat.[172] Indeed, one exercise participant said as much, explaining that when the exercise directors informed the troops that the Blue Force had inflicted personnel casualties and equipment damages, they "were faced with a 'lost situation' from the start and could not just follow the protocol. You have to respond with what you have truly got and that's what a real battle is about."[173] At the end of the exercise, officers and soldiers were praised for carrying out an exercise that was "true, difficult, strict, and actual" ("真难严实"), as well as for "not using a script" (不拿稿子) and "not being afraid to make mistakes" (不怕问倒).[174]

The prominent role given to joint long-range precision strikes in implementing the dictates of target-centric warfare is best illustrated by a CCTV "Live News" broadcast describing a force-on-force confrontation that lasted a little over an hour:

> The joint ground-and-air firepower strike group spread out over a mountainous area of several hundred square kilometers: Armed helicopters skimmed over the ground, assault planes and fighter jets darted past high in the sky, and various artillery batteries of large calibers launched firepower assaults and precision strikes on important targets along the enemy's forward defense positions as well as those in the enemy's depths. . . . As the operations picked up steam, the joint ground-and-air firepower strike group executed targeted clearing of the remaining enemy targets. From the air and on the ground, long-range artillery and armored vehicle guns constituted a firepower strike system that featured combination and meticulous coordination of long- and short-range firepower.[175]

The coordination of firepower mentioned here was accomplished in part by another innovation, described in further detail in the previous section on Mission Action-2013: the shift from traditional force structures organized by unit to mixed combined groups (混合编组) organized by task.[176] In this case, however, reports did not clarify whether these new groups incorporated troops from different services.

---

[171] "Military Report" [军事报道], CCTV-7, November 2, 2013c.

[172] Mei Changwei and Li Dezhong, 2013

[173] Liang Shenhu [梁申虎], Fu Xiaohui [付晓辉], and Mei Changwei [梅常伟], "For This Battle, We Fight from a 'Lost Situation' – Account on Jinan Military Region's "Queshan Decisive Victory-2013" Real Troop Exercise" [这一丈，从"残局"开打 — 济南军区"确山决胜—2013"实兵演习见闻], *Liberation Army Daily* [解放军报], November 9, 2013, p. 2.

[174] Li Dezhong and Ma Yongsheng, 2013.

[175] "Live News" [新闻直播], CCTV Xinwen, November 4, 2013b.

[176] Li Dezhong and Ma Yongsheng, 2013.

"Night Tigers" with "New Fangs": Joint Decisive Victory-2013

Unsurprisingly, reconnaissance and joint precision strikes were equally important features of Joint Decisive Victory-2013, which consisted of two different types of joint nighttime operations: a tri-service joint beach seizing and landing focused on joint intelligence gathering, maritime transport, assault landing, and joint firepower strikes and (like Queshan Decisive Victory-2013A) a joint mountain offensive operation focused on continuous target-centric precision strikes and integration of information and firepower.[177]

One aspect of target-centric warfare that appeared to receive particular attention during this exercise was the third tenet of the concept—regulating or controlling the nature and timing of operations based on the extent to which a given target has been destroyed. While previous daytime joint exercises had practiced "changing control of the course of the battle" away from proceeding according to scripted time points and toward proceeding based on an assessment of damage to a target (目标毁伤效果), this exercise apparently saw debate among "grassroots-level commanders" as to the applicability and relevance of this approach in nighttime operations.[178] Nevertheless, a *Liberation Army Daily* account of the exercise described a specific instance in which a concealed amphibious assault group waited for its signal to move into battle as an air attack group used firepower strikes to hit key enemy targets overhead; the members of the amphibious assault group knew that the commanders would give the order to engage "at the last minute based on the effect the air attack achieved."[179]

Just as in Queshan Decisive Victory-2013A, Joint Decisive Victory-2013 featured a stronger Blue Force designed to simulate actual combat and challenge the Red Force to deal with unexpected situations. At one point, as the Red Force beach landing formation attempted to traverse the sea, "the Blue Force relentlessly carried out electronic surveillance and jamming, launched disruptive attacks in the air, and staged a blockade at sea, posing major threats to the Red Force."[180] Reflecting on this aspect of the exercise, Jiang Jianbo, the military training chief of the North Sea Fleet headquarters, remarked on the crucial role that the PLAAF and other aviation troops play in cross-sea transport, stating, "During the course of maritime ferrying, the main thing is coordination with the Air Force and Navy aviation for aerial cover and protection."[181]

---

[177] "Military Report" [军事报道], CCTV-7, November 19, 2013e.

[178] Fu Xiaohui [付晓辉], Wang Weidong [王卫东], and Yue Lixin [岳立新], "Tactical Research and Training, Real Knives and Real Guns, Striking the Opponent Dead with One Blow: On-the-Scene Report from Jinan Military Region's Nighttime Live-Forces Exercise 'Joint Decisive Victory-2013,' Part II" [战法研练，真刀真枪练就"一剑封喉" — 济南军区"联合决胜-2013"夜间实兵演习新闻观察之二], *Liberation Army Daily* [解放军报], December 6, 2013.

[179] Fu Xiaohui, Wang Weidong, and Yue Lixin, 2013.

[180] "Military Report" [军事报道], CCTV-7, November 18, 2013d.

[181] "Military Report," 2013e.

The PLAAF also received special recognition for its unique utility in nighttime reconnaissance. A *Liberation Army Daily* article described a moment during the exercise when a team of infantry reconnaissance troops, artillery reconnaissance troops, and special forces found themselves suddenly enveloped in darkness and unable to see their enemy target several kilometers away. It was then that "the joint campaign large formation command center ordered an Air Force electronic reconnaissance plane to quickly determine the information about the target and immediately direct air and ground firepower in a precision strike against the enemy."[182] The lesson here, according to Xue Zhengyun, the commander of the joint campaign large formation command center, was clear: "When the Army can't see there is still the Air Force, and when the Air Force can't see there are still satellites. If you don't think in those terms, then you can't talk about victory on the night battlefield!"[183]

Yet Joint Decisive Victory-2013 also revealed numerous problems, which was perhaps to be expected given the novelty of the exercise. An air defense brigade of the Jinan MR's 26th Group Army, for example, noted that this was the first time it had participated in nighttime tri-service joint beach seizing and landing and thus the first time it had dealt with the subjects of intelligence reconnaissance and early warning, joint firepower coordination, joint defense against air raids, and joint command and control.[184] Adding to the troops' inexperience with these important concepts was the challenge of building a joint air defense system to cover multiple air spaces during the night. After witnessing his troops' performance during the exercise, Xie Feng, the commander of the air defense brigade, remarked that "the coordination in nighttime ground-and-air joint defense against air raids was particularly poor on some occasions."[185] In general, the exercise exposed multiple shortcomings and areas for improvement in nighttime joint operations and target-centric warfare, including "a lack of smooth information-sharing channels, a lack of precision in joint firepower strikes, a delay in target damage assessment, and weak joint counter-jamming capabilities."[186]

Conclusion

Joint operations and the closely associated concepts of informatization and system-of-systems warfare have played a significant role in shaping the PLA's modernization efforts over the past two and a half decades. These ideas have influenced the development of platforms,

---

[182] Mei Changwei [梅常伟], Wang Huifu [王会甫], and Yue Lixin [岳立新], "The Night Battlefield, to Be Victorious Requires a String of Pearls and Jade Even More: On-the-Scene Report from Jinan Military Region's Nighttime Live-Forces Exercise 'Joint Decisive Victory-2013,' Part III" [夜战场，制胜更需珠联璧合 一 济南军区"联合决胜-2013"夜间实兵演习新闻观察之三], *Liberation Army Daily* [解放军报], December 7, 2013.

[183] Mei Changwei, Wang Huifu, and Yue Lixin, 2013.

[184] "Military Report," 2013g.

[185] Military Report," 2013g.

[186] "Military Report," 2013e.

munitions, sensors and the operational concepts that tie these technologies into an integrated whole. Through the PLA's operational research and experimentation process, the PLAAF has attempted to build these capabilities into an overarching architecture capable of confronting the United States—its most sophisticated competitor—and its allies. The exercises discussed in the preceding pages represent a limited portion of the numerous PLA training activities conducted over the past decade that tested, refined, and demonstrated new modes of warfighting and training. Since the early 2000s, the PLA has used these same training events as a means of advancing comprehensive training reform efforts intended to institutionalize more-realistic training and exercises that more closely approximate the increasingly complex combat environment PLAAF operators will encounter in future wars. Taken at face value, these exercises and training reform efforts suggest major improvements in PLAAF capabilities. In several cases, however, these reforms have failed to address persistent problems that severely limit the PLAAF's ability to improve in several key areas.

Press reports and journal articles indicate that early experimentation efforts left many in the PLA fully aware of the significant distance still required to develop actual capabilities that could be used in combat.[187] These exercises provided PLA and PLAAF leaders with an understanding of their shortcomings and of the need for practical experience in planning and executing more-realistic training, as well as an appreciation of the complexity of these new forms of warfare. It is unclear, however, whether subsequent training events have successfully implemented these lessons in any meaningful way. Although the more recent exercises highlighted in the preceding pages suggest some level of improvement and increased sophistication, the military press reports that accompanied them discuss many of the same problems that surfaced in Sharp Sword 2005—difficulty in coordination, obstacles to information sharing, limited realism, and a continued reliance on scripting.

These shortfalls have not gone unnoticed and generally are not glossed over in the after-action reports and media accounts published in PLA military science journals and news periodicals. Despite the relative lack of information and the political messages behind many of the articles published in these outlets, the authors—including the midlevel and senior officers who frequently write or are quoted in the articles—have demonstrated an awareness of these problems and a relatively consistent openness in acknowledging their existence. Nevertheless, repeated references to new programs, initiatives, and reforms attempting to address these shortfalls provide evidence that problem recognition is less of an obstacle than finding effective solutions and implementing them.

Many in the PLA and PLAAF continue to question the quality and realism of their training and understand that these shortcomings greatly impair their actual wartime capability. The exercises examined here highlight the PLAAF's role in the new informatized warfare for which

---

[187] "An Expedition That Spans History," *Battle Flag News* [战旗报], March 9, 2006; and Wang Jianmin, "Footprints of the Forerunner," *Battle Flag News* [战旗报], February 16, 2006.

PRC and PLA leaders have been pushing their commanders to prepare. In many cases, the concepts that underpin these experimentation and training events signify a broad understanding of the types of functions that PLAAF units will be expected to perform in any future conflict scenario. Beyond the concepts, however, it remains unclear just how sophisticated PLAAF capabilities have become. Limited planning experience and problems with information sharing and coordination indicate that key functions, such as air defense, targeting, battle damage assessment, and air support to ground forces, will remain challenges until the PLAAF can address these systemic problems. Absent significant changes that provide for more-realistic and complex training, the PLAAF's capabilities will continue to be constrained by its outdated organizational culture.

References

"An Expedition That Spans History,"[1] *Battle Flag News* [战旗报], March 9, 2006.

Allen, Kenneth W., "Introduction to the PLA's Administrative and Operational Structure," in James C. Mulvenon and Andrew N. D. Yang, eds., *The People's Liberation Army as Organization*, Santa Monica, Calif.: RAND Corporation, CF-182-NSRD, 2002, p. 1–44. As of July 13, 2016:
http://www.rand.org/pubs/conf_proceedings/CF182.html

Chang Xin [常欣], "Use Military Innovation Theory to Guide Substantial Leap in Combat Power" [以军事创新理论牵引战斗力实质跃升], *Vanguard News* [前卫报], March 31, 2013.

Chen Youyuan [陈有元], "Deepen Training in the New 'Three Attacks and Three Defenses'" [深化新'三打三防'训练], *Liberation Army Daily* [解放军报], November 14, 2000.

Cheng Sixun, "Exploration and Practice of Integrated Training of Military Region Units: Part One," *Battle Flag News* [战旗报], February 9, 2006a.

———, "Exploration and Practice of Integrated Training of Military Region Units: Part Two," *Battle Flag News* [战旗报], February 10, 2006b.

———, "Exploration and Practice of Integrated Training of Military Region Units: Part Three," *Battle Flag News* [战旗报], February 14, 2006c.

Ding Yahan [丁雅涵] and Li Dezhong [李德忠], "In This Battle, Aiming at the 'Vulnerable Spot' When Firing—Close-Up View of the Military Region's 'Queshan Decisive Victory-2013A' Exercises with Troops, Part III" [这一丈，瞄准"七寸"打—近观军区"确山决胜－2013A"实兵演习之三], *Vanguard News* [前卫报], December 8, 2013.

Finkelstein, David, "China's National Military Strategy: An Overview of the 'Military Strategic Guidelines,'" in Roy Kamphausen and Andrew Scobell, eds., *Right-Sizing the People's Liberation Army: Exploring the Contours of China's Military*, Carlisle, Pa.: Institute for Strategic Studies, May 2007, pp. 82–87.

Fu Quanyou, "Deepening Military Training by Scientific and Technological Means," Xinhua, June 27, 2000.

Fu Xiaohui [付晓辉], Wang Weidong [王卫东], and Yue Lixin [岳立新], "Tactical Research and Training, Real Knives and Real Guns, Striking the Opponent Dead with One Blow: On-the-Scene Report from Jinan Military Region's Nighttime Live-Forces Exercise 'Joint Decisive Victory-2013,' Part II" [战法研练，真刀真枪练就"一剑封喉" － 济南军区"联合

决胜-2013”夜间实兵演习新闻观察之二], *Liberation Army Daily* [解放军报], December 6, 2013.

Ge Dongsheng, ed., *On National Security Strategy,* Beijing: Military Science Publishing House, 2006.

Han Guozhang [韩国章], "Inspiration of and Reflection on 'Mission-Action-2010A' Exercise" ['使命行动－2010A'演习的启示与思考], *Comrades-in-Arms News* [战友报], November 2, 2010.

Hao Yuqing and Cai Renzhao, eds., *The Science of Armed Forces Building*, Beijing: National Defense University Press, 2007.

Huang Bin, *Research into the Kosovo War,* Beijing: Liberation Army Publishing House, 2000.

Information Office of the PRC State Council [中华人民共和国国务院新闻办公室], "Defense White Paper: The Diversified Employment of China's Armed Forces" [国防白皮书：中国武装力量的多样化运用], April 16, 2013.

Kang Yongsheng [康永升] and Zhang Dianfu [张殿福], "The 'Mission Action' Military Exercise Has No 'Deep-Level Intentions'" ["使命行动"军演没有"深层用意"], *China Youth News* [中国青年报], September 27, 2013.

Li Dezhong [李德忠] and Ma Yongsheng [马永生], "The 'Queshan Juesheng-2013A' Live Personnel Exercise Concludes Very Satisfactorily: Deepening Utilization of the Military Region Unit Training Reform and Innovation Achievements" ["确山决胜－2013A"实兵演习圆满落幕：深化运用军区部队训练改革创新成果], *Vanguard News* [前卫报], November 10, 2013.

Li Yongsheng, *Lectures on the Science of Joint Campaigns,* Beijing: Military Science Press, 2012.

Liang Hongqing [梁洪清], "Integrated Command Platform Employed in Large-Scale Real Combat for the First Time" [一体化指挥平台首次大规模投入实战应用], *Battle Flag News* [战旗报], October 21, 2010.

Liang Shenhu [梁申虎], Fu Xiaohui [付晓辉], and Mei Changwei [梅常伟], "For This Battle, We Fight from a 'Lost Situation'—Account on Jinan Military Region's 'Queshan Decisive Victory-2013' Real Troop Exercise" [这一丈，从"残局"开打—济南军区"确山决胜－2013"实兵演习见闻], *Liberation Army Daily* [解放军报], November 9, 2013.

Lin Hongguan, Yang Ping, and Liu Quan, "Assault in the Air and on the Ground—Account of the Participation of a Certain Aviation Division of the Beijing Air Force in the 'Mission Action-2010' Exercise," *Air Force News* [空军报], October 25, 2010.

Liu Fengan [刘逢安] and Wu Tianmin [武天敏], "New-Generation 'Outline of Military Training and Evaluation' Issued" [新一代《军事训练预考核大纲》颁发], *Liberation Army Daily* [解放军报], July 24, 2008.

———, "Training Transformation: Striving to Advance Amid Wind and Rain – Interview with Chen Zhaohai, Director of Military Training and Service Arms Department of PLA General Staff Department" [训练转变：在风雨中奋力前行—访总参军训和兵种部部长陈照海], *Liberation Army Daily* [解放军报], November 2, 2010.

Liu Shenyang [刘沈扬], "The Theory and Practice of Target-Centric Warfare" [目标中心战的理论与实践], *China Military Science* [中国军事科学], No. 5, 2013, pp. 83–92.

Liu Yuejun, ed., *Non-Contact Warfare,* Beijing: Military Science Publishing House, 2004.

"Live News" [新闻直播], CCTV Xinwen, October 9, 2010.

"Live News" [新闻直播], CCTV Xinwen, October 16, 2013a.

"Live News" [新闻直播], CCTV Xinwen, November 4, 2013b.

"Live News" [新闻直播], CCTV Xinwen, November 17, 2013c.

Lu Feng and Ni Menzhi,"Mobile and Camouflaged Launches Using New Equipment Under Complex Weather and Terrain Conditions," *People's Front* [人民前线], July 28, 2004.

Mei Changwei [梅常伟] and Li Dezhong [李德忠], "Exercise Starts This Time with a 'Scenario of Losing Battle'—Second Installment of Report Series on the Queshan Decisive Victory-2013A Exercises in the Jinan MR" [这一仗，从"残局"开打—近观军区"确山决胜－2013A"实兵演习之二], *Vanguard News* [前卫报], December 3, 2013.

Mei Changwei [梅常伟], Wang Huifu [王会甫], and Yue Lixin [岳立新], "The Night Battlefield, to Be Victorious Requires a String of Pearls and Jade Even More: On-the-Scene Report from Jinan Military Region's Nighttime Live-Forces Exercise 'Joint Decisive Victory-2013,' Part III" [夜战场，制胜更需珠联璧合—济南军区"联合决胜-2013"夜间实兵演习新闻观察之三], *Liberation Army Daily* [解放军报], December 7, 2013.

"Military Report" [军事报道], CCTV-7, October 16, 2010.

"Military Report" [军事报道], CCTV-7, September 22, 2013a.

"Military Report" [军事报道], CCTV-7, September 27, 2013b.

"Military Report" [军事报道], CCTV-7, November 2, 2013c.

"Military Report" [军事报道], CCTV-7, November 18, 2013d.

"Military Report" [军事报道], CCTV-7, November 19, 2013e.

"Military Report" [军事报道], CCTV-7, November 25, 2013f.

"Military Report" [军事报道], CCTV-7, November 30, 2013g.

Mou Dong [牟东], Zhou Shiai [周世艾], and Liu Chang [刘畅], "For the First Time, the Military Region Proposes and Actually Constructs and Uses a Shaped Network Command Information System" [军区首次提出并实际构建运用栅格化网络指挥信息系统], *Soldiers News* [战士报], October 27, 2013.

Mulvenon, James C., and Andrew N. D. Yang, eds., *A Poverty of Riches: New Challenges and Opportunities in PLA Research,* Santa Monica, Calif.: RAND Corporation, CF-189-NSRD, 2004. As of July 15, 2016:
http://www.rand.org/pubs/conf_proceedings/CF189.html

Pan Youmu, *The Study of Non-Contact Warfare,* Beijing: National Defense University Publishing House, 2003.

Peng Guangqian and Yao Youzhi, eds., *The Science of Military Strategy,* Beijing: Military Science Publishing House, 2005.

Qing Yong and Mo Fuchun, "First Phase of Trans-Region Mobility Exercise 'Mission Action-2010B' Completed, Command Information System Subjected to Network Attack Throughout Entire Course," *People's Army* [人民军队], October 20, 2010.

Ren Chaohai [任朝海], "Accomplish the Transformation of Theoretical Results into Work Results" [实现理论成果向工作成果转化], *Liberation Army Daily* [解放军报], November 22, 2000.

Shao Guopei, ed., *Information Operations in Non-Contact Wars,* Beijing: Liberation Army Publishing House, 2004.

Song Linsheng [宋林胜], "Target Centric Warfare, a Deadly System-Breaking Weapon" [破击作战体系的利器 — 目标中心战], *China Military Online* [中国军网], February 6, 2013.

Song Youfa and Hong Yaobin, eds., *Integrated Joint Operations Command Headquarters Work,* Beijing: Military Science Press, 2005.

Sun Jie [孙杰] and Fu Xiaohui [付晓辉], "Queshan Juesheng 2013A Exercise Enters Real-Troop, Live Fire Combat Stage" ["确山决胜-2013A"演习进入实兵实弹战斗阶段], *China National Radio* [中国广播网], November 4, 2013.

Wang Houqing and Zhang Xingye, eds., *The Science of Campaigns,* Beijing: National Defense University, 2001.

Wang Jianmin, "Footprints of the Forerunner," *Battle Flag News* [战旗报], February 16, 2006.

Wang Wendui [王文锥] and Le Hongwang [骆宏望], "Troop Projection in 'Packages,' No 'Going Separate Ways'" ["打包投送"拒绝"各走各路"], *Soldiers News* [战士报], October 16, 2013.

Wang Yongming, Liu Xiaoli, and Xiao Yunhua, *Research into the Iraq War,* Beijing: Military Science Publishing House, 2003.

Yang Qingmin, "'Mission Action-2010A' Trans-Region Mobile Exercise of the Beijing Military Region Shows Three Major Highlighting Features" [使命行动 2010A 北京军区跨区机动演习凸显三大亮点], Xinhua, October 21, 2010.

Xinhua [新华] "PLA Begins 'Mission Action-2013' Exercise on the Tenth" [解放军 10 日"使命行动－2013"演习], September 9, 2013.

Xiong Weiwei, "Seriously Implement the Series of Important Decisions and Instructions of Chairman Xi and Making Persistent Efforts in Deeply Pushing Forward Actual Combat-Oriented Training" [认真贯彻习主席系列重要决策指示再接再厉深入推进实战化训练], *Air Force News* [空军报], October 27, 2014.

Xu Sheng [徐升], "Panorama Talk on Science and Technology Military Exercises" [科技练兵纵横谈], *Liberation Army Daily* [解放军报], August 10, 1999.

Xue Xinglin [薛兴林], ed., *Campaign Theory Study Guide* [战役理论学习指南], Beijing: National Defense University Press [国防大学出版社], 2001.

Yang Hong [杨鸿], "Military Region Conducts Exercise 'Mission Action-2010A' Oath Swearing and Mobilization Rally" [军区举行'使命行动－2010A'演习誓师动员大会], *Comrades-in-Arms News* [战友报], October 12, 2010.

Zhan Yu [战玉], "A Study of the Theory of Integrated Joint Operations" [一体化联合作战理论探要], *China Military Science* [中国军事科学], No. 6, 2007, pp. 11–21.

Zhang Fenghai [张凤海] and Zhou Feng [周峰], "Successful Conclusion of Trans-Regional Mobile Campaign Exercise 'Mission Action-2013A'" ["使命行动－2013A"跨区机动战役演习圆满落幕], *People's Front* [人民前线], September 20, 2013a.

———, "The Important Platform in the Military Strengthening Practice—Several Points of Revelation and Thought About the 'Mission Action-2013A' Cross-Region Mobile Campaign Exercise," *People's Front* [人民前线], October 18, 2013b.

Zhang Li [张力], Wu Aili [武艾丽], and Zhao Lingyu [赵凌宇], "System-of-Systems Operation Directs Sword at Distant Sea—Witnessing the Air Force's Trans-Regional Maneuver Campaign Exercise 'Mission Action-2013C'" [提子作战剑指远海—空军"使命行动—2013C"跨区机动战役演习闻思录], *Air Force News* [空军报], November 1, 2013.

Zhang Li [张力] and Zhu Qianli [朱谦礼], "Air Force Units Fulfill Staged Tasks in the Trans-Region Mobile Exercise; Precision Command Drives Ground-Air Joint Operations" [空军部队在跨区机动演习中圆满完成阶段性任务—精确指挥牵引陆空联合作战], *Liberation Army Daily* [解放军报], October 27, 2010.

Zhang Wenping and Yan Wenbo, "Advance Phase of Second Artillery's Integrated Training Starts—Establishing Steering Group on Integrated Training, Organizing Trial Comprehensive Integration and Integrated Training, Conducting Theoretical Study on Integrated Combat and Training, and Exploring Characteristics and Laws of Integrated Training," *Rocket Forces News* [火箭兵报], July 13, 2004.

Zhang Yuliang [张玉良], ed., *The Science of Campaigns* [战役学], Beijing: National Defense University Press [国防大学出版社], 2006.

Zhuang Lijun [庄利军], "A Rapidly Expanding Transformation in the Training Domain" [训练领域一场方兴未艾的变革], *Liberation Army Daily* [解放军报], February 6, 2006.

Chapter Three

New Type Support

Developments in PLAAF Air Station Logistics and Maintenance

Timothy R. Heath

Summary

The People's Liberation Army (PLA) Air Force (PLAAF) has undertaken extensive reforms of its support services as part of a broader transformation as a "strategic air force." The latest reforms aim to build a logistics and maintenance system capable of supporting a diverse array of sustained offensive and defensive operations from anywhere in the country. Developing a system of comprehensive support bases lies at the center of this effort, but reforms also include the streamlining of command and control and the integration of information networks. The PLAAF is also revising maintenance and supply procedures and improving rapid deployment capabilities to better support the higher operational demands. Despite some progress, the development of what China calls a "new type" support system remains in an early stage. Formidable obstacles remain, including poor integration of new technology, slow progress in the modernization of air bases, and inadequate numbers of trained personnel.

Introduction

The U.S. Air Force is presently considering options to improve operational resilience through some combination of adjustments to its overseas basing structure, including dispersion, hardening, and enhanced repair and defensive capabilities. While main operating bases are unlikely to be abandoned, they are likely to be supplemented by a network of more-austere bases. The PLAAF has shown a similar growing appreciation of the importance of operational resilience. It has begun to develop a system of secondary, or reserve, bases capable of supporting dissimilar aircraft types. However, China's basing overhaul remains primarily focused on the creation of a handful of major air bases capable of serving a broad range of operational and training functions, a concept similar to the U.S. Air Force main operating base. The acquisition of Russian SU-27 (Chinese J-11) and Chinese J-10 and J-H7 aircraft at the turn of the century, in particular, has spurred a spate of construction and renovation activities aimed at developing the facilities, equipment, and maintenance and logistics technology to support the new platforms. The PLAAF's vision of a "strategic air force" is also driving reforms in communications, information networks, maintenance and logistics procedures, and other changes necessary to support the vision of aviation units capable of executing a broad variety of sustained offensive operations across the breadth of Chinese territory.

Background: The PLAAF's Transformation into a "Strategic Air Force"

In 2004, Central Military Commission (CMC) Chair Hu Jintao elevated the importance of the development of capabilities to shape a favorable security environment as well as better protect China's core interests when he proclaimed the "historic missions of the armed forces."[188] Through this mission set, the central leadership directed the military to prevail in traditional combat against adversaries across the breadth of China's territorial or sovereignty claims, including the maritime regions of the East and South China Seas, remote mountainous western border regions, and Taiwan. It also increased the demand on the PLA to execute military operations other than war, including citizen evacuation abroad, humanitarian aid/disaster relief at home and abroad, military-to-military engagement, and participation in United Nations peacekeeping operations.

In response to these new requirements, the PLAAF articulated a vision of itself as a "strategic air force" defined by the capabilities to "ensure information dominance," carry out "integrated air and space operations," execute "both offensive and defensive operations simultaneously," and "respond throughout the entire territory."[189] An article in the *People's Daily* explained that the heart of this transformation lies in "establishing an information system" supported by "space-

---

[188] James Mulvenon, "Chairman Hu and the PLA's New Historic Missions," *China Leadership Monitor*, No. 27, Winter 2009.

[189] "From Supportive Service to Strategic Air Force: Major Change in China's Air Force Buildup Thinking," *Hong Kong Feng Huang Wang*, June 28, 2004.

based information platforms" that can "horizontally cover all positions on the basis of vertical ground-air-space integration" and serve as an "integrated network at the strategic, operational, and tactical levels."

The information system is envisioned as a sort of brain and nervous system for a "new type operational power system" characterized by "air-space integration, offensive-defensive integration, information firepower integration, and operations-support integration." The PLAAF envisions the new type of operational power system as integrating "strategic early warning, long-range strike, air defense and antimissile operations, information offense and defense, and strategic power projection." This vision underpins all of the PLAAF's modernization efforts since at least 2005.[190]

The PLAAF Embraces "New Type Support"

The organizations most directly affected by the call for "new type support" are the PLAAF Logistics Department and its Equipment Department. The PLA's dictionary of terms defines the term "support" as "organized implementation of support and service type activities required for a military unit to carry out its responsibilities and fulfill other requirements. According to the mission requirements, this may be further divided into combat support, logistics support, and equipment support." The dictionary further defines "military logistics" as the "provision by national or military authorities of all materials required to support war fighting or military construction. This includes finance; general supplies; petroleum, oil, and lubricants (POL); sanitation; transportation; and military facilities construction."[191] Accordingly, the PLAAF Logistics Department's primary responsibility concerns the provision of supplies for construction, operations, training, and daily life. The Logistics Department has 18 subordinate departments, bureaus, divisions, and offices responsible for various aspects of the overall system. These include Headquarters, Politics, Finance, Quartermaster, Health, Equipment, Transportation, Fuels, Materials, Airfield Construction, Barracks Management, Air Materiel, and others. Each of these offices is represented through the chain of command from the PLAAF Logistics Department all the way down to the lowest unit level. The Logistics Department is also responsible for the maintenance of equipment belonging to antiaircraft artillery, surface-to-air missiles, radar, communications, or airborne troops.

The PLAAF Equipment Department is responsible for determining how much and what types of equipment should be procured and for general management, maintenance, repair, and procurement of aircraft and aircraft ground support equipment. The Chinese military dictionary

---

[190] Party's Innovation Theory Study and Research Center of the PLAAF, "Direct the Modernization Building of the Air Force According to the Goal of Military Strengthening" [以强军目标统领空军现代化建设], *People's Daily* [人民日报] November 18, 2013.

[191] *China's People's Liberation Army Military Dictionary* [中国人民解放军军语], 军军语 Academy of Military Science Publications: Beijing, China, 2011, pp. 66, 475.

of terms defines "aircraft maintenance," a major responsibility of the PLAAF Equipment Department, as the "application of technical equipment and procedures to ensure the flying status of aircraft." Aircraft maintenance includes "daily maintenance, periodic maintenance, and special day maintenance."[192] However, the Equipment Department also carries out research, development, testing, and evaluation of equipment. The PLAAF Equipment Department predated the General Equipment Department (GED), which was established as a fourth General Department in 1998. Since then, however, the PLAAF's Equipment Department reports to the GED and has adopted its terminology. The PLAAF has at least 21 repair factories to carry out major repairs and aircraft and engine overhauls. Aviation units have repair factories, which are responsible for intermediate and minor repairs. Each flight group (飞行大队) has a dedicated maintenance squadron (机务大队) assigned to it. The maintenance squadrons are responsible for receiving, managing, and maintaining airframes throughout their service life.[193] At the troop level, the logistics and maintenance systems come together in the form of support teams. Each airframe, for example, has a dedicated ground crew that includes both logistics and maintenance team personnel.[194]

PLAAF logistics and maintenance forces have undergone several major reforms since their founding as "supply units" (供应部) in the 1950s. In the PLAAF's first decades, the primary responsibility of these units centered on dedicated support to flight missions, a role also known as flight services (飞行服务). In 1979, however, the responsibilities of the PLAAF logistics units expanded to include greater care for personnel and equipment. The PLAAF issued its first extensive set of regulations on logistics support in 1991 (空军后勤保障条例), which was further modified following the release of new military strategic guidelines in 1993.[195] Incremental reforms in the 2000s focused on improving the ability of the logistics and maintenance systems to provide support in near-combat conditions. This included efforts to improve support to nighttime and dissimilar aircraft operations, rapid runway repair, and the development of a joint logistics system.[196]

Since 2005, the pursuit of the "new type of combat power" has profoundly shaped the PLAAF's approach to its logistics and maintenance systems. Momentum to meet the requirements of the strategic air force concept accelerated in the late 2000s. In 2009, the annual PLAAF Logistics Work Conference outlined directives for a series of "new type operations

---

[192] *China's People's Liberation Army Military Dictionary* [中国人民解放军军语], 2011, p. 1013.

[193] Allen, Kenneth, "PLA Air Force Logistics and Maintenance: What Has Changed?" in James C. Mulvenon and Richard R. Yang, eds., *The People's Liberation Army in the Information Age*, RAND Corporation, CF-145-CAPP/AF, 1999, pp. 80–82.

[194] James C. Mulvenon and Andrew N. D. Yang, eds., *The People's Liberation Army as Organization*, Santa Monica, Calif.: RAND Corporation, CF-182-NSRD, 2002, pp. 272–299.

[195] Xinhua, "PLAAF Air Base Equipment Maintenance Breakthroughs" [中国空军机场配备新型装备破解排弹抢修难题], October 27, 2009.

[196] Allen, 1999, p. 84.

logistics support systems," including the "multitype aircraft comprehensive support base," "multidirectional simultaneous full territory support system," and "large-sized high-intensity sustainable support" to support the "strategic transformation of the PLAAF."[197]

At the first forum on the development of modern logistics, held on August 17–18, 2010, the PLAAF reportedly drew up a "blueprint" for the "future development of the PLAAF's logistics." It provided direction on PLAAF logistics training and reviewed proposals to reform logistics to support the PLAAF's growing operational demands.[198] That year, General Logistics Department director Liao Xilong affirmed that the "common goal of all military logistics is to basically complete the task of comprehensive construction of modern logistics" by 2020.[199] In 2011, a PLA-wide work conference on "Accelerating Logistics Construction" summed up relevant "pilot" reforms and directed a military-wide implementation of work to modernize the logistics systems.[200] In 2011, the CMC, the four General Departments, and the PLAAF Party Committee issued the "Outline on Building Modern Logistics in a Comprehensive Way" (全面建设现代后勤纲要), which outlined in greater detail the way forward for reforming PLAAF logistics and maintenance services. The outline identified new tasks to be achieved over the coming years, including the elaboration of a system of air bases, the construction and integration of command networks, the strengthening of highly mobile, modular logistics forces, and the improvement of training and recruitment of skilled personnel.[201]

The PLAAF equipment maintenance system has demonstrated a similar focus on reform to better support the vision of a strategic air force. At a PLAAF-wide work conference on equipment held in May 2012, PLAAF Commander Xu Qiliang outlined broad guidance for reforms to the equipment and maintenance work of the PLAAF to be realized in the 12th Five-Year Program. He called for the "reform of the support model and mechanisms," the "reliance on new technologies and new means to implement condition-oriented maintenance," and changes in maintenance procedures to be focused on the "individual aircraft's performance status." He also directed the "scientific adjustment" of the "assignments of equipment service support tasks."[202]

---

[197] Zhang Jinyu and Li Jianwen, "PLAAF New Type Operation Logistics Support Systems Established," *Jiefangjun Bao Online* in English, January 7, 2009.

[198] "PLA Air Force to Attain Epoch-Making Progress in Aviation Maintenance Support Model," *Jiefangjun Bao Online* in English, August 19, 2010.

[199] Fan Juwei and Zhang Jinyu, "Promote Innovative Development in Comprehensive Construction of Modern Logistics from a New Starting Point" [在新起点上推进全面建设现代后勤创新发展], *PLA Daily* [解放军报], November 19, 2010.

[200] Wang Shibin and Fan Juwei, "Guo Boxiong Attends PLA Work Conference on Accelerating Logistics Construction," *Jiefangjun Bao Online* in English, December 1, 2011.

[201] Gong Dean, Xu Chunxiang, and Tian Yuanlong, "Giant Strides Taken in Logistics Construction in New Period" [新时期后勤建设步伐], *Air Force News* [空军报], November 25, 2011.

[202] Zhou Zhaoxia, "PLAAF Equipment Work Conference Held in Beijing" [空军装备工作会议在北京举行], *Air Force News* [空军报], May 22, 2012.

The following section reviews aspects of PLAAF logistics and maintenance developments in line with these directives.

Air Base System Construction

A major focus of these reforms has been on the physical infrastructure of air bases to support offensive and defensive operations. Changes have centered on developing a network of differentiated facilities capable of providing resilient, rapid, and sustained support to air operations of diverse aircraft types occurring throughout the breadth of Chinese territory. The changes mark a significant increase in demand on PLAAF airfields and the logistics units that serve at the facilities.

The PLAAF's military airfields, many built in the 1950s and 1960s, have proved ill equipped to support contemporary needs. Throughout the late 1990s and 2000s, the PLAAF updated the airfields to accommodate new platforms. The acquisition of Russian SU-27 (Chinese J-11) and Chinese J-10 and J-H7 aircraft at the turn of the century, in particular, spurred a spate of construction and airport renovation activity to accommodate the new platforms. In 2003, for example, the Lanzhou Military Region Air Force (MRAF) reportedly carried out renovation and repair of numerous airfields, POL depots, military rails, and roads to accommodate new aircraft. The MRAF Logistics Department also overhauled electricity supplies, lighting systems for night flight, and navigation systems.[203]

Similarly, for years the PLAAF has sought to improve the resilience and survivability of its airfields. It began studying rapid repair for damaged aircraft and airfields after the first Gulf War. During the early 2000s, the PLAAF began creating a rapid-repair subunit at each airfield and conducting training for "on-site rapid repair" of equipment that had been subjected to enemy attack. Each subunit incorporates personnel from the maintenance group and air station. The PLAAF has expanded the focus of relevant drills to include dissimilar aircraft. A 2005 drill on emergency response and rapid repair held in Fujian, for example, featured "10 models of aircraft and 30 types of fighter planes."[204]

Field logistics units have historically been assigned to an "air station" (场站), which is an organization composed of all logistics troops directly responsible for flight support duties and assigned to an airfield (机场). The air station is typically a regimental-grade command subordinate to the aviation unit it serves. Air station responsibilities include aviation materiel, POL, ordnance supply, maintenance and protection of the airfield, security, communications, navigation, weather, and transport.[205] As of 2002, the PLAAF classified its air stations according

---

[203] Zhang Jinyu, "Lanzhou MAC Air Force Upgrades Its Logistics Support Capability," *Jiefangjun Bao Online* in English, November 27, 2003.

[204] Song Fang, "Nanjing Air Force Units' Three in Depth Preparations for Military Battle" [南空部队军事斗争备从深行], *Air Force News* [空军报], June 14, 2005.

[205] *China's People's Liberation Army Military Dictionary* [中国人民解放军军语], 2011, p. 958.

to their ability to support flight operations. Grade A (甲) air stations could simultaneously support up to two aviation regiments, Grade B (乙) could support one regiment, Grade C (丙) air stations could support airfields in caretaker (i.e., not currently in use but could be upgraded and reactivated) (看守) or alternate (备降) status, while Grade D (丁) air stations could serve only airfields in caretaker status.[206]

Despite incremental improvements to the capabilities of air stations, the PLAAF Logistics Department concluded in 2007 that its approach to airfields needed to be rethought. The PLAAF has moved toward an "air base" model in which, as much as possible, one facility can "support multiple deployment locations, multiple aircraft types, and different aircraft models in operations."[207] This capability would facilitate the movement of fighter, strike, and other aircraft across military regions to defend against air attack, strike enemy ground forces, transport troops, and carry out other military operations anywhere in China. In November 2008, a PLAAF conference directed modifications to air stations to better support the PLAAF's "progress towards air base transformation" (向基地化发展). The term *air force base* is a facility at which logistics troops are permanently stationed, but it also serves as a key node for training and operations. At the same time, the air stations continue to maintain all of their traditional responsibilities, including airport maintenance and flight services, POL, aviation material storage, general military supply, air field maintenance and repair, communications, weather analysis, and large-scale air transportation. However, the execution of these tasks is modified and updated to better support the offensive operations of the aviation units.[208] The conceptual shift from airfields toward air bases is meant to signal a shift in mentality better attuned to the needs of the PLAAF as a strategic air force.

After years of research, the PLAAF articulated its blueprint for the development of a system of air stations adapted to the air base concept in the 2011 "Outline for Building Modern Logistics in a Comprehensive Way." The outline prioritized the construction of comprehensive support air bases characterized as "operational entities, support bases, and the nodes of combat systems." It also directed the "systematic consideration and planning for the development" of the air stations to operate on the air bases.[209] In January 2014, the sixth PLAAF meeting on air station construction mapped out a blueprint for the construction of air bases. It featured directions on the

---

[206] Liu Daoqin, *Science of Air Force Tactical Logistics* [空军指挥学院], Air Force Command Academy Press: Beijing, 2002, p. 34.

[207] Liu Fazhong, "Vigorously Exploring a New Base for Aviation Unit Logistics Support Model" [积极探索航空兵后勤保障基地化新模式], *PLA Daily* [解放军报], December 22, 2007.

[208] Xinhua, 2009.

[209] Gong Dean, Xu Chunxiang, and Tian Yuanlong, 2011.

construction of air base facilities; the development of training, operations, and support functions; and the development of air station units accordingly.[210]

The 2011 outline and 2014 plan for air base construction feature a system of four major types of bases: (1) comprehensive support bases (综合保障基地), regarded as the "operational hubs"; (2) multi–airplane-type support bases (多机种保障基地), regarded as the "backbone"; (3) basic task support bases (基本任务保障基地); and (4) reserve support bases (备用保障基地). The latter two types of bases are envisioned as "serving supplementary roles." The 2011 outline directed that comprehensive support base functions be "improved" by 2015, with "some" air bases meeting the requirement for serving multiple models of aircraft. Moreover, a number of comprehensive support bases and a "large number" of multi–airplane-type support bases are expected to be operational by 2020.[211] All four types of bases are expected to develop certain standard features. They are expected to carry out construction and renovation of facilities to support the new requirements, standardize logistics and maintenance procedures, and install information networks. However, each type of base features its own unique characteristics, reviewed below.

Comprehensive Support Bases

As the cornerstone of the air base system, the comprehensive support base features the most extensive support capabilities of any PLAAF facility and serves a primary role in managing operations. A 2009 PLAAF book explained that these "core air bases" feature robust command, logistics, and equipment support and maintenance capabilities. The comprehensive support bases serve as a "command, communication, and intelligence center, a base for supporting the training and operation of multiple types of aircrafts, and a distribution center for operations-related material." [212]

The comprehensive support base's primary function is to serve as the foundation for nationwide air operations, both defensive and offensive. These facilities also provide support to the main types of combat, transport, and special mission aircraft that the PLAAF operates. In addition, they provide radiating support (辐射支援) to auxiliary bases in the neighboring areas (周边机场) and support to other nearby ground-based forces (地面部队保障能力).[213]

Some features of the comprehensive support base can be discerned from available media reporting. First, the bases are large facilities featuring at least one large resident air unit, probably a regiment, which is subordinate either to a tenant air division headquarters at the airfield or to an air brigade. The comprehensive support base is also a relatively new phenomenon. The Chengdu

---

[210] Zhang Zhihong and Xiong Shenhui, "Let Warrior Hawks Soar High and Far" [为了战鹰高飞远航], *Air Force News* [空军报], March 20, 2014.

[211] Zhang Zhihong and Xiong Shenhui, 2014.

[212] Zhu Hui, ed., *Strategic Air Force Theory* [战略空军轮], Lantian Publishing, Beijing, China, 2009, p. 98.

[213] Xinhua, 2009.

MRAF built the first comprehensive support base in 2010.[214] As of 2014, only a small number of comprehensive support base construction pilot projects had been completed.[215] While Chinese media have not provided the identities of the bases, the most likely candidates are bases with either brigade or division headquarters that host China's third-generation (U.S. Air Force fourth-generation) aircraft.

Second, the bases share with the multi–aircraft-type base extensive information networks (discussed below). These bases also feature a "three-level" command structure, which consists of an MRAF-level command center, the air base, and the control tower. The air base also features information networks that connect the various logistics and maintenance functions into a "support command center."

Third, the bases feature capabilities designed to improve survivability and resilience. A typical base features special-purpose support equipment, such as a vehicle to supply provisions for pilots, runway sweepers, runway rapid repair vehicles, and navigational light vehicles. At one such base, detachments[216] developed "multifunctional refueling vehicles" which allow one refueling vehicle to refuel many different models of fighter planes. The base also carried stocks of operational and training materials, such as ammunition, petroleum, aviation materiel, field camping equipment, and operational training material stores. This allowed various types of aircraft to return to the base and perform maintenance and replenishment.[217]

Fourth, logistics and maintenance units on the bases regularly carry out training to improve combat readiness. Air Force newspapers routinely describe training events, including mock air raids and drills in putting out fires, defusing unexploded ordnance, restoring runway surfaces and power supply, and laying out emergency navigation lights. One base established a flight support-training facility covering 18 specialties so that troops could conduct training and learn new trades without leaving the base.[218]

Multi–Aircraft-Type Support Bases

These support bases are primarily focused on providing support to multiple aircraft types, both resident and deployed. In early 2007, the PLAAF held its first conference on constructing multi–aircraft-type comprehensive support bases. In September of that year, tens of different models of aircraft successfully took off and landed in a Nanjing Military Region (MR) airfield, which PLAAF press hailed as the "first multiple type aircraft comprehensive support base." The key feature of these bases is their ability to support dissimilar aircraft. One Nanjing MRAF air

---

[214] Gong Dean, Xu Chunxiang, and Tian Yuanlong, 2011.

[215] Zhang Zhihong and Xiong Shenhui, 2014.

[216] "Detachments" [分队] are battalion and below ad hoc organizations, which are also called elements, subunits, and grassroots organizations.

[217] Liu Fazhong, 2007.

[218] Huo Chao, "Striving to Increase Cross-Model Aircraft Support Ability" [有力提升跨机型保障能力], *PLA Daily* [解放军报], February 18, 2013.

base reported that it can now "carry out support tasks for 16 types of military aircraft simultaneously."[219]

Reserve Air Stations

The development of reserve air stations represents a relatively recent development. A reserve air station established in 2005 has supported a single type of aircraft since 2006. It was reportedly "one of the first" of the batch of reserve air stations established by the PLAAF.[220]

The PLAAF has repurposed and reinvigorated some of its smaller air stations into reserve air stations. Because the PLAAF has reduced overall numbers of airplanes as part of its overall modernization program, demand for the quantity of air stations has declined. In 2009, the PLAAF held a meeting to discuss the management of air stations in caretaker status. That year, the air force relabeled these facilities as "reserve air stations." One article explained that the change in terminology reflected the conclusion that such facilities could play a larger role in supporting operational training and become integrated into the air base network. In this manner, the reserve air stations serve as strategic reserve facilities.[221] As with the other two types of air bases, the support units based at reserve air stations also carry out training to improve readiness. For example, a Fujian Province reserve airfield carried out training for flight support and rapid runway repair in 2010.[222]

The reserve air stations also play a role in mobilization. An article published in 2011 explained how a Chengdu MRAF reserve airfield carried out its role in emergency response and mobilization support for the MRAF. It also served as a joint tactical training and a backup landing place for a certain aviation regiment. In these roles, the air station reportedly supported 6,000 sorties over a span of three years.[223] The preparation of civilian airports for wartime roles augments the backup capability represented by the reserve air stations.

Basic Task Bases

Reportedly numbering in the "hundreds," basic task bases likely include older air stations occupied by regiments with a single type of older airplane. They also include facilities that host supply warehouses, ammunition stores, or fuel depots. The basic task bases can also be facilities for PLAAF ground units, such as air defense, radar, and other units. A network of "reserve

---

[219] Chen Tao and Sun Peng, "Integration of Diversified Command Methods Promotes Support Efficiency" [多样化指挥手段融合促进保障高效], KJB *Air Force News* [空军报], March 5 2010.

[220] Ba Jianmin and Xiong Huaming, "Reserve Airfield Station Possesses Emergent Combat Support Capability," *Jiefangjun Bao Online* in English, August 27, 2009.

[221] Xinhua, 2009.

[222] "Reserve Airfield and Station Conducts Drill," *Jiefangjun Bao Online* in English, October 29, 2010.

[223] Zhang Hengping, "The 'Frontline' Thinking of a Reserve Air Station" [一个备用场站的一线思维], *Air Force News* [空军报], March 2, 2011.

warehouses" also stores logistics and maintenance materiel, particularly reserves of ammunition for offensive air operations, as well as aircraft equipment and spare parts.[224]

Information Networks for Support Functions

If the network of air bases provides the physical backbone infrastructure for the new type of support, the command and information networks provide its "nervous system." The 2011 outline highlighted the importance of logistics and maintenance functions in the PLAAF's information network. It established an "informatization office" that would oversee the planning and installation of networks. The goal was to ensure that the "multipurpose support network" could be completed in two years' time and "cover all types of facilities in the Air Force." The outline acknowledged that as of 2011, "substantial progress" had been made and these efforts should be expanded to include "wireless broadband, handheld terminal smart analysis, automatic control, etc." It also directed the application of new technologies, such as "materiel network management and automatic control, sensors for airfield oil tanks, and integrated airport security support automatic surveillance and control systems to control alien objects, birds, obstacles, power supply, and lights."[225]

The PLAAF appears to have made substantial progress in the installation of information networks. Numerous reports in PLAAF media in the 2007–2012 time frame suggest the construction of information networks had advanced enough that some logistics units no longer needed to rely on traditional methods of communication, such as two-way radios, telephones, and intercoms, to pass orders and submit requests for services.[226] The information networks aim to link MRs with subordinate units in the same system, as well as integrate units of different specialties within the same units. For example, a 2011 report stated that the Guangzhou MRAF Equipment Department invested 3 million yuan ($485,000) to set up military regionwide communications networks that connected all subordinate units.[227] An individual air station in the same MR, meanwhile, invested over 1 million yuan ($160,000) to upgrade and convert the existing information system to develop an integrated command platform between various combat and support functions, including intelligence, logistics, maintenance, and operations.[228] Across MRAFs, the main features of the support command networks appear to consist of the following elements.

---

[224] Zhu Hui, 2009, p. 98.

[225] Gong Dean, Xu Chunxiang, and Tian Yuanlong, 2011.

[226] Bi Yanxiang, "Shift from a Dispersed and Simplistic Support System Toward an Integrated and Synthetic Support System" [保障体系：由分散单一向一体合成转变], *Air Force News* [空军报], October 18, 2007.

[227] Zhang Shijie and Tong Xiaojun, "Optimizing Logistics Efficiency" [追求保障效益大直], *Air Force News* [空军报], March 11, 2011.

[228] Xu Guoyong, "A Certain Air Station in the Guangzhou Military Region Sees Great Leap in Its Support Capability" [广空莫场站多兵几种保障大幅提高], *Air Force News* [空军报], April 27, 2012.

Support Command Network

The support command network provides overall command and control of both logistics and maintenance functions and is directly linked to the operations networks. The support command network features four functions of command and control, information integration, safety monitoring, and control of subordinate logistics and maintenance units. Installation of a video monitoring and radio intercom system allows commanders to have a "real-time grasp of the airfield support dynamics," allowing "precision command and control of individual troops and vehicles, more refined support, and the resolution of problems." This contrasts with traditional practices, in which commanders could not see the actual logistics and maintenance work during flight operations and in which orders frequently were not passed down.[229]

Petroleum, Oil, and Lubricant Network

POL detachments operate as part of the air stations, providing direct support to resident air units. However, POL detachments have access to nearby storage facilities, or depots, to replenish their stores. POL depots also enable reserve airfields to serve as redundant facilities in the event of combat. For ground vehicles, PLAAF personnel may access joint logistics POL fueling stations located on the military facilities of other services. To improve the reliability and accountability of its POL resources, logistics units have developed networks connecting refueling stations and storage facilities. For aviation fuel, the installation of fiber-optic cable allows air units to connect distant depots, with some located as far as "100 km away." The networks also facilitate access to POL for ground vehicles for all services, as well as remote, automatic testing of the quality of oil.[230]

Ammunition Information Network

Management of ordnance is divided between logistics and equipment forces. The air stations manage ordnance for antiaircraft artillery and surface-to-air missiles, as well as for paratroopers. However, the maintenance units manage air-launched missiles. Both the logistics and maintenance detachments responsible for ordnance have developed information networks to track munitions stocks and provide rapid, precise replenishment. In the 2007–2008 time frame, PLAAF newspaper reports carried accounts of numerous air stations that had developed "ammunition support information networks," usually developed by the airfield units themselves.

---

[229] Tian Yuanlong, "Actual Account of the PLAAF Logistics Department Building a Multi-Type Aircraft Support Base" [空军后勤部建设多机种保障基地记实], *PLA Daily* [解放军报], December 25, 2011.

[230] "Logistics Sub-Department Conducts Close to Real War Logistics Training," *PLA Daily Online* in English [解放军报], January 14, 2009.

These networks enable personnel to check stocks remotely and automatically, instead of through physical inspection.[231]

Maintenance Network

Maintenance groups have increased their use of information technology to assist with diagnosis, repair, and routine checkups. Maintenance units have developed their own networks that reportedly display "in real time" data on work regarding an airplane's flight parameters and the condition of airframes, engines, and so on. Chinese press also highlights the use of handheld diagnostic equipment to "quickly download flight parameters." The technician then sends the data to the diagnostic center to conduct system analysis. The handheld device is also loaded with "supplementary decision making system software" and data on spare parts and troubleshooting.[232]

Parts and Aviation Supplies Network

At the troop level, the logistics and maintenance systems come together in the form of support teams.[233] These support teams manage the parts and aviation supplies networks. Media reports have claimed that by 2009, an aviation supplies support network had been built for the "entire Air Force." The network reportedly relies on the Chinese global positioning "Beidou" satellite system and features wireless access. Work to standardize and integrate data has traditionally proven a major challenge.[234] Online supply catalogs can feature over "10,000 types of materials and items" for aviation units.[235]

While the PLAAF appears to have completed installation of command and subordinate maintenance and logistics networks for aviation units, depots and other smaller facilities may still be carrying out work to develop and integrate their own local area networks. A 2012 report stated, for example, that "100 percent" of the Beijing MRAF's subordinate aviation units had been connected to the military's comprehensive information network. However, it noted that 78 percent of airfield fuel depots had completed construction of their respective local area networks.[236]

---

[231] Chi Yuguang, Hu Ren, and Yan Guoyou, "A Division of the Air Force Relies on Information Technology to Increase Its Support Capacity," [航空兵某师依靠信息技术提高保障能力], *PLA Daily*, [解放军报], September 15, 2008, p. 3.

[232] Yu Lei, "Equipment Support Command: Change from Traditional Methods Toward Network-Centric Linkage" [保障指挥：由传统方式向网路链接转变], *Air Force News* [空军报], October 25, 2007.

[233] Mulvenon and Yang, 2000, pp. 272–299.

[234] Guo Ziyu, "Advance All-Round Progress of Air Force Aviation Supplies Informatization" [全面推进空军航材信息化建设进程], *Air Force News* [空军报], September 15, 2009.

[235] Chi Yuguang, Hu Ren, and Yan Guoyou, 2008.

[236] Hu Zhuying, "Beijing Military Air Force Logistics System's Informatization: Exploration and Practice" [信息快车：提速后勤保障], *Air Force News* [空军报], February 8, 2012.

The establishment of information networks has facilitated changes in how logistics units carry out support. At the division and regiment levels, logistics staffs have set up "logistics operations offices" designed to coordinate requests between maneuvering units and nearby logistics facilities. However, the PLAAF continues to refine procedures and work processes to make better use of its information networks. In 2014, the PLAAF issued a "Standard Procedure for Informatization of Flight Support for PLAAF Air Stations."[237]

Maintenance Procedure Reforms

The establishment of information networks has also facilitated reform of maintenance procedures. In recent years, the PLAAF has carried out major reforms to the processes and procedures by which it carries out maintenance, especially for its more capable, technologically advanced aircraft. The PLAAF carried out trial reforms from 2010 to 2011, with an unidentified Nanjing MRAF air unit serving as the test unit. The reforms consist of four main elements: (1) customized aircraft maintenance plans, (2) maintenance based on flight hours of aircraft, (3) adjustments to the division of labor, and (4) revamped quality control.[238]

Customized Maintenance Plans

Featuring a high degree of input from technicians, this reform aims to develop customized maintenance plans for individual aircraft. POL and logistics support has shifted to a greater focus on tailored support as well. Media report that support teams are organized around types of airplanes that work in rotating shifts before undergoing final inspection by an expert supervisor. Also characteristic of the changes is a focus on practices that can be sustained when deployed, such as "forward planning" of general support, and the standardization of procedures, rules, regulations, training requirements, and evaluation criteria.[239]

Flexible Schedules

The second major reform concerns the scheduling of maintenance. Traditionally driven by the calendar, the PLAAF has for decades carried out "maintenance days" twice a year. Occurring once in spring and once in fall, these maintenance down-days force air operations to a halt throughout the PLAAF. The reform seeks instead to group airplanes by flight hours to allow a more flexible approach to maintenance. In 2011, the test bed Nanjing MRAF air division stopped "maintenance days" and instead modified maintenance to accommodate individual planes. The

---

[237] Zhang Zhihong and Xiong Shenhui, 2014.

[238] Liu Shenghui, "Four Transformations: Erecting a Generation of New Ladders to Success" [四个转变：架起一代新天梯], *Air Force News* [空军报], May 12, 2010.

[239] Meng Qingbao, "A Shenyang Military Region Air Force Aviation Regiment Finds Solutions for Support Problems While Deployed" [模块化配置解决导地保障难题], *Air Force News* [空军报], November 18, 2011.

division brought together logistics and maintenance teams with scientific research institutes to carry out tailored, comprehensive support.[240]

Maintenance Team Reorganization

The third reform has reorganized maintenance teams into two broad groups. A "flight line" consists of personnel dedicated to carrying out basic tasks to clear and test an aircraft for operations. A "workshop" team, comprising the most-skilled personnel, specializes in the inspection and other more technically challenging maintenance tasks. According to reports from the test bed unit, the "flight line" consists of a contingent of flight clearance technicians and personnel. Such a team enables a single person to complete "power-on checks," preparations, flight clearance tasks, and other work of a more routine nature. This development contrasts with common practices in the past, which reportedly took "five specialized technicians to accomplish" what one person can now do. The report claimed that during routine training, takeoff preparations require less than half the personnel common in traditional practice.[241] Three specialties—ordnance, special installations, and wireless facilities—have been consolidated into a single specialty for "equipment onboard aircraft." These individuals are also responsible for flight preparations and checks of the relevant equipment.[242] The "workshop" team consists of more-senior, experienced expert technicians with access to more-sophisticated diagnostic equipment. These individuals work in a "backstage" environment to carry out research, troubleshooting, and other technical work to maintain the equipment.[243]

Quality Control

A fourth major reform aims to improve the effectiveness, precision, and reliability of quality control. In traditional practice, technicians responsible for carrying out maintenance work also performed inspections for quality. PLAAF media acknowledge problems in this system, the most prominent of which was that maintenance officials preferred to "cover up" problems rather than fix them. The pilot project established the PLAAF's "first independent safety monitoring section." The inspection team, operating separately from the maintenance teams, developed monitoring procedures, standards for quality inspections, rules, regulations, and trial work operations. The quality inspection teams also issued certifications to personnel that passed

---

[240] Li Guowen and Yan Guoyou, "In Training Support, a Certain Nanjing MRAF Air Division Bids Farewell to Collective Maintenance" [南空某航空兵师训练保障告别集体维护], *PLA Daily* [解放军报], June 1, 2011, p. 5.

[241] Liu Shenghui, 2010.

[242] Wang Xuegang and Chao Xiaoming, "An Eyewitness Account of an Exercise for Comparing the Modes of Servicing, Repairing, and Providing Support" [空军莫飞院模型飞机维修保障对比演练现场见闻], *Air Force News* [空军报], October 5, 2010.

[243] Liu Shenghui, 2010.

technical evaluations.[244] The reforms established a "maintenance management office" and a "safety supervision office" in maintenance units to improve quality assurance.[245]

Following the completion of the trial efforts in 2011, the PLAAF directed servicewide implementation of the maintenance reforms in the "Logistics Outline."

Beyond the Air Base: Mobile Logistics Support Teams

The PLAAF's efforts to transform itself into a "strategic air force" have highlighted the importance of integrating operations and support and of enabling the rapid deployment of dissimilar aircraft to distant locations. It has also levied requirements to improve the resilience and survivability of airfields, as well as to ensure the ability of aviation units to carry out operations in combat conditions. The PLAAF's support services must therefore develop the capacity to support the deployment of air assets to any location throughout the country, no matter how remote. PLAAF logistics and maintenance forces have responded by increasing training in rapid deployment and deployment over long distances and by exercising the ability to support the diversion of aircraft to alternate and reserve facilities.

Beginning in the 2000s, the PLAAF stepped up efforts to increase the mobility and emergency responsiveness of its support services. In 2005, for example, units in the Nanjing MR developed a field petroleum pipeline. The pipeline extended from a depot to an airfield "100 km away" and crossed "rivers, streams, and high mountains."[246] By 2006, the PLAAF had established deployable support teams involving over 100 various support vehicles, including command and navigation vehicles, tractors, and fire trucks. The PLAAF conducted exercises in which such teams traveled to an alternate airfield and opened an emergency command post and supply point.[247]

The PLAAF has also sought to increase its ability to support the rapid deployment of flight operations to virtually anywhere in the country. The integration of some functions through the expansion of joint logistics has facilitated this effort. The PLAAF has also increased the use of information technologies to increase the speed, precision, and accuracy of field logistics deployments.

---

[244] Liu Shenghui, 2010.

[245] Li Pengfei and Wang Yue, "A Certain Regiment of the Nanjing Military Region Air Force Passes Acceptance for Maintenance Mode Reform Test" [南空航空兵莫团通过维修改革试点验收], *Air Force News* [空军报], December 15, 2011.

[246] Song Fang, 2005.

[247] Chen Huaping, "Chronicle of Jinan Military Region Air Force Field Station's Organization of Drill for Field Operations Support" [济空莫场站组织野战伴随保障演练记实], *Air Force News* [空军报], December 7, 2006.

Rapid Deployment

The PLAAF continues to develop the capability to deploy support teams to reserve and other alternate airfields. It has also explored additional venues for diversionary airfields, including civilian airports and highways. Civilian construction of major roadways has, in recent years, incorporated military standards into consideration while under construction. In 2014, for example, a Jinan MR air station organized a test for dissimilar aircraft training on a freeway in eastern Henan Province for the first time. The freeway's design supported the takeoff and landing of third-generation combat aircraft.[248]

Long-Distance Deployments

PLAAF support services have stepped up training in long-distance deployments. In the exercise "Iron Flow-2011," the PLAAF carried out its first large-scale maneuver of entire units involving hundreds of vehicles and more than 1,000 people over long distances. The drill tested the ability of the PLAAF to move large amounts of materiel across vast distances, with a convoy from the Lanzhou MR traveling to the Beijing MR. The action reportedly lasted 41 days and involved 11 automotive detachments, 1,157 people, 417 vehicles, 126 rail cars, and two large-sized roll-on/roll-off ships. Each vehicle traveled 3,669 km, rail mileage reached 9,080 km, and sea transport distance totaled 38 nautical miles. The convoy delivered 14,375 tons of aviation to POL units along the way. The exercise also tested satellite navigation and video transmission systems.[249]

Joint Logistics

PLA efforts to improve joint logistics also facilitate the PLAAF's ability to deploy forces throughout the country. Following a successful two-year trial in the Jinan MR, Hu Jintao approved the "implementation plan" for joint logistics to be applied to the "entire military" in December 2007.[250] By 2011, the MRs had indigenously developed database systems for joint logistics support. The MR-level Joint Logistics Departments built a single entity joint logistics support network that included a logistics support demand database covering common equipment and supplies. This has facilitated access to medical services, fuel, provisions, and general supplies to proximate deployed units regardless of service affiliation.[251] According to one 2012 report, the reforms have changed the former arrangement in which each unit's logistics

---

[248] Li Tiegang, "PLA Air Force Successfully Conducts First Dissimilar Aircraft Takeoff and Landing Drill on Freeway" [我空军首次成功举行多机种高速公路起降], *China Air Force Online* [中国空军网], May 26, 2014.

[249] Xu Weijun, "Action in Iron Flow 11" [铁流－11 在行动], KJB *Air Force News* [空军报], July 6, 2011.

[250] Liao Xilong, "Personally Experiencing Jinan Military Region's Major Joint Logistics Reforms" [亲历济南战区大联勤改革], *PLA Daily* [解放军报], December 16, 2008.

[251] Yao Xinyan and Yang Yabin, "Joint Logistics of Three Services Expand Toward Future Battlefield" [三军连勤向未来战场拓展], *PLA Daily* [解放军报], December 28, 2011.

organization managed its own support. Instead, deployed units can now draw food, fuel, and basic supplies from whichever logistics unit happens to be closest.[252]

Information Technology

PRC media report that PLAAF logistics units are incorporating more information technologies in emergency support training and exercises. In a 2014 exercise, for example, a Lanzhou MRAF logistics unit retrofitted support vehicles with onboard wireless terminals and Beidou Global Positioning System (GPS) and provided personnel with handheld terminals.[253]

Case Study: Chengdu Military Region Air Force Dual-Runway Exercise, 2012

The PLAAF remains in the initial stages of developing the support capabilities required to enable a more flexible, resilient, and potentially expeditionary air power. However, glimpses of the service's ambitions can be seen in a handful of exercises that test relevant concepts and capabilities. An exercise held on November 29, 2012, featured sustained support to a wide variety of airframes at a relatively high sortie rate. Operating simultaneously from dual runways, support assets at the unidentified Chengdu MRAF comprehensive support base provided flight services for 12 aircraft types, including fighters, bombers, transports, and early warning aircraft. During the exercise, which lasted five hours and 28 minutes, 85 airplanes took off and touched down at a "high frequency."

In the exercise, the base operated according to the wartime command model (战时指挥模式), featuring a "three level command structure" (三级指挥机构): Chengdu MRAF command center (成都军区空军指挥中心), air base (基地), and control tower (塔台). The exercise illustrated the different responsibilities of each level in the command structure. The Chengdu MRAF command post exercised remote command and control of the flight operations via an automated command information system. The air base command center controlled the support functions, as well as activities related to the operation of the air base facility. The control tower, meanwhile, controlled the takeoff and landing of aircraft. The configuration of the airfield consisted of a dual-runway system that allowed for a larger number of simultaneous sorties.

The air base command center featured an aircraft maintenance and support center (机务保障中心), which exercised command and control of logistics and maintenance functions. An automated command information system allowed the command center to monitor the progress of flight services and make adjustments in a timely manner. In one instance, one of the refueling stations nearly ran out of fuel. The maintenance and support center immediately directed the

---

[252] Hu Zhuying, 2012.

[253] Zhang Junjie, "Smoothly Paving the Way for Combat Aircraft to Soar" [铺平战鹰冲天路], *Air Force News* [空军报], January 24, 2013.

logistics command center (后勤指挥中心) to provide a refueling truck to the location to replenish the station. The maintenance and support center also directed an "aircraft maintenance command center" (机务指挥中心) to provide maintenance and flight preparations support. The aircraft maintenance command center featured an "aircraft preparation status management system" that displayed the status of aircraft parking and preparation. It used a "stoplight chart of red, yellow, green" to signal which planes had malfunctions and which were ready for takeoff.

On the ground, logistics teams grouped by type of aircraft provided tailored support to planes as they landed. In ten minutes, 12 sorties reportedly came and went without interruption. Support services had been organized in functionally defined teams to ensure the orderly and precise execution of tasks. These included five relatively fixed support teams on the two sides of the dual runways with standardized and comprehensive support modules tailored to specific aircraft types.[254]

The exercise demonstrated how improvements in logistics and maintenance procedures, air base configuration, and use of information technologies are expanding capabilities to operate at a higher operational tempo. Therefore, it may be seen as a trial exercise for which the PLAAF is likely to draw lessons on how to improve the ability of its support services to sustain high sortie rates in combat and other missions. Chengdu MRAF Commander Fang Dianrong stated that the exercise marked the "first step in the PLAAF's transformation of operations support modes towards base-centric comprehensive support."[255]

An Incomplete Transformation

The Chengdu MRAF exercise provides a glimpse of the direction that the PLAAF aims toward. However, this event remains an unusual exception to the general practice of logistics and maintenance, which remains defined by the persistence of legacy practices, habits, and equipment. An assessment by the Innovative Theory Research Center of the PLAAF acknowledged in 2013 that the "comprehensive support system is still relatively backward."[256] PLAAF writings note a variety of problems plaguing the support forces, including the following.

Poor Integration of Information Technology

Despite progress in basic installation and dissemination, the PLAAF's support forces continue to have problems with integrating information technologies. A 2014 PLAAF newspaper article noted that the overall efficacy of efforts by air stations to incorporate information

---

[254] Cao Yinan and Hu Xiaoyu, "Forge New Operation Training Support Platform" [打造战训保障新平台], *Air Force News* [空军报], December 7, 2012.

[255] Liu Jun and Sun Yang, "Echelons of Combat Aircraft Strike Out Roaring" [鹰阵绽风雷], PLA Pictorial [解放军画报], December 16, 2012.

[256] Party's Innovation Theory Study and Research Center of the PLA Air Force, 2013.

technology remained unsatisfactory. It emphasized problems in the connection and integration of communications and information technology and noted that a standardized mechanism to operate the technologies had not been built.[257] Similarly, the Nanjing MRAF Equipment Department has observed that, "some units" achieved "substantial results in the building of information systems" but "did not convert what they had achieved into support power on the battlefield." It stated that the units "did not link the building of information systems with training support and actual operations requirements."[258]

Inefficient Deployment Capability

Despite exercises that demonstrated progress in extending the distance and responsiveness of rapid deployment logistics and maintenance teams, the overall ability of the PLAAF to field support services remains limited. A PLAAF assessment described the ability of air bases to deploy support teams as "weak." It called common practices inefficient and wasteful, highlighting "clumsy and awkward support services" that involve "excessive manpower and material resources."[259]

Lack of Professional Culture

PLAAF press criticizes the persistence of outdated practices and a tendency in the support services to lack a sense of professional pride and attention to detail. A 2014 PLAAF newspaper article criticized the persistence of "conventional concepts and forms" of support services, which it claimed continued to play a "dominant role."[260] Another article criticized formalistic adherence to new procedures without actual implementation. It warned, "having a system but not implementing it does more harm than not having a system." The article argued for the development of a professional culture. It prescribed measures to promote an environment in which the implementation of new processes and procedures is made into a "self-aware, voluntary consensus of all the officers and troops." Among methods adopted, one air station printed and issued information and teaching materials and held speech contests and knowledge competitions.[261]

---

[257] Zhang Zhihong and Xiong Shenhui, 2014.

[258] Liang Yansheng, "Use Information Technology to Raise the Effectiveness of Support Equipment" [运用信息化装备保障能力提升], *Air Force News* [空军报], July 3, 2012.

[259] Zhang Zhihong and Xiong Shenhui, 2014.

[260] Zhang Zhihong and Xiong Shenhui, 2014.

[261] Yao Binbin and Yang Wenhui, "Journal on the Enhancement of Training Quality and Results" [鹰击长空而不逾短], *Air Force News* [空军报], June 11, 2013.

Insufficient Numbers of Trained Personnel

The limited numbers of personnel with advanced technical skills and training remain a persistent constraint on the PLAAF support services. The military has struggled to compete with the more lucrative civilian sector for individuals with the technical backgrounds and education needed to ensure precise and accurate maintenance of advanced platforms. The 2011 Modern Logistics outline accordingly identified the cultivation of experts as a priority.

Insufficient Fuel and Supplies

PLAAF units continue to experience restrictions on fuel use. Fuel availability, along with maintenance time on engines and airframes, remains one of the driving factors that determine the annual flying quota for units and individual pilots. To address the shortage of parts and equipment for some aircraft, the PLAAF continues to impose restrictions on flight operations. One 2010 PLAAF article noted problems, including an "acute lack of POL sources," an "uneven layout of refineries," and "irregular POL distribution." It also noted that insufficient quantities of jet fuel for more-advanced aircraft lead to stoppages in refinery production that pose a substantial threat to sustainment of flight operations.[262]

Persistence of Corruption

A sensitive topic in official military press, the persistence of corruption in the logistics services nevertheless remains a serious constraint on the ability of the support services to meet the needs of operational units. The Xi Jinping administration has stepped up national-level efforts to curb corrupt practices throughout the military, including the PLAAF.[263]

Conclusion: Reforms Remain in an Early Stage

The PLAAF remains in the early stages of a major transformation of its support forces to better enable the rapid deployment and sustained execution of a diverse array of offensive and defensive air operations across the breadth of the nation. If achieved, this capability could facilitate the movement of fighter, strike, and other aircraft across MRs to defend against air attack, carry out strikes against enemy ground forces, or support military operations anywhere across Chinese territory and its border regions. The PLAAF has made substantial progress in establishing the basic infrastructure of bases and information networks to enable this vision. However, the number of comprehensive support bases remains small and inadequate for the PLAAF's requirements. The 2012 Chengdu exercise, hailed as a cutting-edge demonstration of

---

[262] Han Yunlong and Feng Junjie, "Ensuring That Every Drop of Fuel Is Used Effectively" [让每一滴油都有效燃烧], *Air Force News* [空军报], April 29, 2010.

[263] Michael Chase, Jeffrey Engstrom, Tai Ming Cheung, Kristen A. Gunness, Scott Harold, Susan Puska, and Samuel K. Berkowitz, *China's Incomplete Military Transformation: Assessing the Weaknesses of the People's Liberation Army (PLA)*, Santa Monica: Calif.: RAND Corporation, RR-893-USCC, 2015.

PLAAF logistics and maintenance modernization, underscores the weaknesses of the support services. While the U.S. Air Force has operated at high sortie rates on a 24-hour-a-day, seven-day-a-week combat footing for years in the Iraq and Afghanistan wars, only in the past few years has the PLAAF demonstrated an ability to maintain a higher-than-normal sortie rate for a few hours. These shortcomings will even more severely constrain efforts to operate beyond China's periphery at a sustained pace.

Moreover, the persistence of legacy practices and issues, poor integration of information technologies, inadequately trained personnel, and inexperience will continue to impose serious constraints on the ability of the PLAAF to carry out the missions it aspires to as a "strategic air force." The PLAAF only recently abandoned the practice of observing maintenance days that ground the entire air force to a halt twice a year. Modernization efforts have improved the performance of the support services, but also opened new vulnerabilities. The growing use of information networks to manage workflows related to logistics and maintenance, for example, could open PLAAF networks to cyber attacks. Chinese media widely acknowledge problems in cyber security for military information networks.

The PLAAF leadership appears to recognize the imperative of overcoming the limitations of the logistics and maintenance systems. The 2011 Modern Logistics outline, 2014 base construction plan, and other major planning documents point to serious resource commitments by PLAAF leadership to address these shortcomings. The logistics and maintenance services thus will continue to serve as a critical barometer of the PLAAF's transformation into a modern air force.

References

Allen, Kenneth W., "PLA Air Force Logistics and Maintenance: What Has Changed?" in James C. Mulvenon and Richard R. Yang, eds., *The People's Liberation Army in the Information Age*, RAND Corporation, CF-145-CAPP/AF, 1999, p. 79–88. As of July 13, 2016: http://www.rand.org/pubs/conf_proceedings/CF145.html

Ba Jianmin and Xiong Huaming, "Reserve Airfield Station Possesses Emergent Combat Support Capability," *Jiefangjun Bao Online* in English, August 27, 2009.

Bi Yanxiang, "Shift from a Dispersed and Simplistic Support System Toward an Integrated and Synthetic Support System" [保障体系：由分散单一向一体合成转变], *Air Force News* [空军报], October 18, 2007.

Cao Yinan and Hu Xiaoyu, "Forge New Operation Training Support Platform" [打造战训保障新平台], *Air Force News* [空军报], December 7, 2012.

Chase, Michael, Jeffrey Engstrom, Tai Ming Cheung, Kristen A. Gunness, Scott Harold, Susan Puska, and Samuel K. Berkowitz, *China's Incomplete Military Transformation: Assessing the Weaknesses of the People's Liberation Army (PLA)*, Santa Monica: Calif.: RAND Corporation, RR-893-USCC, 2015. As of July 13, 2016: http://www.rand.org/pubs/research_reports/RR893.html

Chen Huaping, "Chronicle of Jinan Military Region Air Force Field Station's Organization of Drill for Field Operations Support" [济空莫场站组织野战伴随保障演练记实], *Air Force News* [空军报], December 7, 2006.

Chen Tao and Sun Peng, "Integration of Diversified Command Methods Promotes Support Efficiency" [多样化指挥手段融合促进保障高效], *Air Force News* [空军报], March 5, 2010.

Chi Yuguang, Hu Ren, and Yan Guoyou, "A Division of the Air Force Relies on Information Technology to Increase Its Support Capacity," [航空兵某师依靠信息技术提高保障能力], *PLA Daily*, [解放军报], September 15, 2008.

China's People's Liberation Army Military Dictionary [中国人民解放军军语], Academy of Military Science Publications: Beijing, China, 2011.

Fan Juwei and Zhang Jinyu, "Promote Innovative Development in Comprehensive Construction of Modern Logistics from a New Starting Point" [在新起点上推进全面建设现代后勤创新发展], *PLA Daily* [解放军报], November 19, 2010.

"From Supportive Service to Strategic Air Force: Major Change in China's Air Force Buildup Thinking," *Hong Kong Feng Huang Wang*, June 28, 2004.

Gong Dean, Xu Chunxiang, and Tian Yuanlong, "Giant Strides Taken in Logistics Construction in New Period" [新时期后勤建设步伐], *Air Force News* [空军报], November 25, 2011.

Guo Ziyu, "Advance All-Round Progress of Air Force Aviation Supplies Informatization" [全面推进空军航材信息化建设进程], *Air Force News* [空军报], September 15, 2009.

Han Yunlong and Feng Junjie, "Ensuring That Every Drop of Fuel Is Used Effectively" [让每一滴油都有效燃烧], *Air Force News* [空军报], April 29, 2010.

Hu Zhuying, "Beijing Military Air Force Logistics System's Informatization: Exploration and Practice" [信息快车：提速后勤保障], *Air Force News* [空军报], February 8, 2012.

Huo Chao, "Striving to Increase Cross-Model Aircraft Support Ability" [有力提升跨机型保障能力], *PLA Daily* [解放军报], February 18, 2013.

Li Guowen and Yan Guoyou, "In Training Support, a Certain Nanjing MRAF Air Division Bids Farewell to Collective Maintenance" [南空某航空兵师训练保障告别集体维护], *PLA Daily* [解放军报], June 1, 2011.

Li Pengfei and Wang Yue, "A Certain Regiment of the Nanjing Military Region Air Force Passes Acceptance for Maintenance Mode Reform Test" [南空航空兵莫团通过维修改革试点验收], *Air Force News* [空军报], December 15, 2011.

Li Tiegang, "PLA Air Force Successfully Conducts First Dissimilar Aircraft Takeoff and Landing Drill on Freeway" [我空军首次成功举行多机种高速公路起降], *China Air Force Online* [中国空军网], May 26, 2014.

Liao Xilong, "Personally Experiencing Jinan Military Region's Major Joint Logistics Reforms" [亲历济南战区大联勤改革], *PLA Daily* [解放军报], December 16, 2008.

Liang Yansheng, "Use Information Technology to Raise the Effectiveness of Support Equipment" [运用信息化装备保障能力提升], *Air Force News* [空军报], July 3, 2012.

Liu Daoqin, *Science of Air Force Tactical Logistics* [空军指挥学院], Air Force Command Academy Press: Beijing, 2002.

Liu Fazhong, "Vigorously Exploring a New Base for Aviation Unit Logistics Support Model" [积极探索航空兵后勤保障基地化新模式], *PLA Daily* [解放军报], December 22, 2007.

Liu Jun and Sun Yang, "Echelons of Combat Aircraft Strike Out Roaring" [鹰阵绽风雷], PLA Pictorial [解放军画报], December 16, 2012.

Liu Shenghui, "Four Transformations: Erecting a Generation of New Ladders to Success" [四个转变：架起一代新天梯], *Air Force News* [空军报], May 12, 2010.

"Logistics Sub-Department Conducts Close to Real War Logistics Training," *PLA Daily Online* in English [解放军报], January 14, 2009.

Meng Qingbao, "A Shenyang Military Region Air Force Aviation Regiment Finds Solutions for Support Problems While Deployed" [模块化配置解决导地保障难题], *Air Force News* [空军报], November 18, 2011.

Mulvenon, James, "Chairman Hu and the PLA's New Historic Missions," *China Leadership Monitor,* No. 27, Winter 2009. As of July 13, 2016: http://media.hoover.org/sites/default/files/documents/CLM27JM.pdf.

Mulvenon, James C. and Andrew N. D. Yang, eds., *The People's Liberation Army as Organization*, Santa Monica, Calif.: RAND Corporation, CF-182-NSRD, 2002. As of July 13, 2016: http://www.rand.org/pubs/conf_proceedings/CF182.html

Party's Innovation Theory Study and Research Center of the PLAAF, "Direct the Modernization Building of the Air Force According to the Goal of Military Strengthening" [以强军目标统领空军现代化建设], *People's Daily* [人民日报] November 18, 2013.

"PLA Air Force to Attain Epoch-Making Progress in Aviation Maintenance Support Model," *Jiefangjun Bao Online* in English, August 19, 2010.

"Reserve Airfield and Station Conducts Drill," *Jiefangjun Bao Online* in English, October 29, 2010.

Song Fang, "Nanjing Air Force Units' Three in Depth Preparations for Military Battle" [南空部队军事斗争备从深行], *Air Force News* [空军报], June 14, 2005.

Tian Yuanlong, "Actual Account of the PLAAF Logistics Department Building a Multi-Type Aircraft Support Base" [空军后勤部建设多机种保障基地记实], *PLA Daily* [解放军报], December 25, 2011.

Wang Shibin and Fan Juwei, "Guo Boxiong Attends PLA Work Conference on Accelerating Logistics Construction," *Jiefangjun Bao Online* in English, December 1, 2011.

Wang Xuegang and Chao Xiaoming, "An Eyewitness Account of an Exercise for Comparing the Modes of Servicing, Repairing, and Providing Support" [空军莫飞院模型飞机维修保障对比演练现场见闻], *Air Force News* [空军报], October 5, 2010.

Xinhua, "PLAAF Air Base Equipment Maintenance Breakthroughs" [中国空军机场配备新型装备破解排弹抢修难题], October 27, 2009. As of June 15, 2016: http://news.xinhuanet.com/mil/2009-10/27/content_12335819.htm

Xu Guoyong, "A Certain Air Station in the Guangzhou Military Region Sees Great Leap in Its Support Capability" [广空莫场站多兵几种保障大幅提高], *Air Force News* [空军报], April 27, 2012.

Xu Weijun, "Action in Iron Flow 11" [铁流－11 在行动], *Air Force News* [空军报], July 6, 2011.

Yao Binbin and Yang Wenhui, "Journal on the Enhancement of Training Quality and Results" [鹰击长空而不逾短], *Air Force News* [空军报], June 11, 2013.

Yao Xinyan and Yang Yabin, "Joint Logistics of Three Services Expand Toward Future Battlefield" [三军连勤向未来战场拓展], *PLA Daily* [解放军报], December 28, 2011.

Yu Lei, "Equipment Support Command: Change from Traditional Methods Toward Network-Centric Linkage" [保障指挥：由传统方式向网路链接转变], *Air Force News* [空军报], October 25, 2007.

Zhang Hengping, "The 'Frontline' Thinking of a Reserve Air Station" [一个备用场站的一线思维], *Air Force News* [空军报], March 2, 2011.

Zhang Jinyu, "Lanzhou MAC Air Force Upgrades Its Logistics Support Capability," *Jiefangjun Bao Online* in English, November 27, 2003.

Zhang Jinyu and Li Jianwen, "PLAAF New Type Operation Logistics Support Systems Established," *Jiefangjun Bao Online* in English, January 7, 2009.

Zhang Junjie, "Smoothly Paving the Way for Combat Aircraft to Soar" [铺平战鹰冲天路], *Air Force News* [空军报], January 24, 2013.

Zhang Shijie and Tong Xiaojun, "Optimizing Logistics Efficiency" [追求保障效益大直], *Air Force News* [空军报], March 11, 2011.

Zhang Zhihong and Xiong Shenhui, "Let Warrior Hawks Soar High and Far" [为了战鹰高飞远航], *Air Force News* [空军报], March 20, 2014.

Zhou Zhaoxia, "PLAAF Equipment Work Conference Held in Beijing" [空军装备工作会议在北京举行], *Air Force News* [空军报], May 22, 2012.

Zhu Hui, ed., *Strategic Air Force Theory* [战略空军轮], Lantian Publishing, Beijing, China, 2009.